MONSTER

Paul Roehrig | Ben Pring

**A Tough Love Letter On Taming
The Machines That Rule Our Jobs,
Lives, and Future**

WILEY

Published by John Wiley & Sons, Inc., Hoboken, New Jersey.
Published simultaneously in Canada.

For general information on our other products and services or for technical support, please contact our Customer Care Department within the United States at (800) 762-2974, outside the United States at (317) 572-3993 or fax (317) 572-4002.

Wiley publishes in a variety of print and electronic formats and by print-on-demand. Some material included with standard print versions of this book may not be included in e-books or in print-on-demand. If this book refers to media such as a CD or DVD that is not included in the version you purchased, you may download this material at http://booksupport.wiley.com. For more information about Wiley products, visit www.wiley.com.

Library of Congress Cataloging-in-Publication Data is available:

ISBN 9781119785910 (Hardcover)
ISBN 9781119785927 (ePDF)
ISBN 9781119785934 (ePub)

Cover art, original illustrations, and interior book design by Andy Barker at theethicalbrandingco.co.uk

SKY10022402_012921

Contents

Intermission:

Sunflower: When tech meets capital

Contents

1 HAVE WE CREATED A MONSTER?

In which we explore one of the great questions of our time: Have we inadvertently created technology that is hurting our society, our economy, and even our minds?

We love technology. Waze, Netflix, Shazam, Hotel Tonight, Spotify, the MRI that diagnosed Ben's back problem, Gmail, Headspace, Alexa, even on occasion the corporate travel application. Technology is a miracle — something that has made billions of people's lives around the world materially better. Including ours.

We — Ben and Paul — have worked in tech most of our professional lives, as IT analysts, management consultants, and technology practitioners, playing a small role in creating and shaping an industry that employs a significant percentage of the world's working population and is now worth an eye-watering $4 trillion a year.[1] We have unashamedly been technology evangelists. But recently something has changed, and now we've become worried. Why? Because we increasingly come across not as tech evangelists but as tech apologists.

> "AI is the great story of our time!" we say (on stages around the world).
> "Data is the new oil."
> "Everything that can be automated will be automated."
> "Hyper-personalization is key to competitive advantage."
> "Don't be a bad robot — be a good human being."
> "Contact tracing is key to stopping the coronavirus."
> "Pre-digital dinosaurs roam the earth. Don't be one."

People nod, and often applaud, which is nice, but then the real questions start.

> "How many jobs will AI destroy?"
> "What should my kids study?"
> "How can we compete against pure digital companies?"
> "What will ordinary, non-tech-savvy people do in a world of brilliant tech superstars?"
> "Will we need to sacrifice our freedom for our health?"
> "How can I beat the robots?"

"What about Universal Basic Income?"
"Will the Fourth Industrial Revolution lead to a real revolution?"
"What scares you?"

Typically, we nod, pause, smile, and say, "That's a very good question." Then we try our best to convey a message that acknowledges the concern in the questioner's mind but also provides a positive, hopeful point of view: "If we take the right actions now, things are going to be OK. Better than OK, in fact."

Lately, though, we've started feeling less certain that things are going to be "better than OK."

And it's in that light that we attempt in this short book to ask and begin to answer perhaps the most important questions of our time:

> Have we inadvertently built some kind of technology monster that is attacking our **society**, our **economy**, and even our individual **psychology**? If so, what should we do about it?

In our last book, *What To Do When Machines Do Everything* (published 2017), we didn't shy away from the impact tech has had — and will have — on disrupting jobs or spurring other downsides of progress and innovation. We laid out a vision that artificial intelligence (AI) and other new technologies are simply the next generation of powerful productivity tools for us to use wisely. These tools will change our world, as new tools always have, by taking us to the next level of potential and achievement.

So far, overall, we've been largely right. While it's true that the pandemic is clearly reshaping how and where we work, it's also the case that forecasts of AI and robots causing a job apocalypse were overblown. Before the pandemic, employment numbers were at record levels in the Western world, and many sectors show signs of quick recovery. Being "pragmatic optimists" about technology has seen us stand out in crowds of doom-mongers and dystopians.

But concerns about technology's negative side have grown stronger and stronger since we published *Machines*, and despite our best efforts to the contrary, the *zeitgeist* that surrounds technology has become steadily bleaker. Even at a moment when tech has been a lifeline for people stuck at home during the COVID-19 lockdowns.

Central to this darkening mood have been four key trends:

1. The persistent sense of dread (even in the absence of any real evidence) that brilliant machines will outpace even the most brilliant of minds.

2. The ubiquity of social media (and growing awareness of its negative impact at a micro and macro level).

3. The unholy *pas de deux* between "big money" and "big tech."

4. The pervasive feeling that in aggregate, tech is making our jobs, personal lives, and even our societies somehow worse, rather than better.

Combined, these dynamics have soured the perception of technology as a force for good, and left many questioning the core tenets of technology's role in our lives and societies.

Including, now, us. We *love* technology, remember? But even we are asking ourselves, "What the hell is going on?" Anyone looking at the daily news — except for the most myopic and naïve — could easily think, "Jeez, we are collectively losing our minds!"

Social conventions of privacy and courtesy are melting away. Our democratic institutions — fair elections, civil discourse — seem as quaint and distant as buggies and gas street lamps. Once cool, disruptive tech "rock stars" are being exposed as nothing more than the latest digital robber barons, propped up by easy money that arrives as an "offer you can't refuse" with few questions attached but in reality is a clear demand to "make me a boatload more money." Increases in aggression, depression, and self-harm are seen by some as signs that our new machines are melting our brains.

Minds, money, machines, society — together, these systems weave a complex web of history, economics, sociology, religion, law, and politics. They are all interconnected, and together they are morphing the rules of our jobs, lives, and societies in a way we haven't seen since the First Industrial Revolution.

We can feel it, and you probably can too. While many good things are happening around us, we *know* we can do better. To improve, though, we must recognize the ground truth about where we are.

These dynamics have, of course, only intensified as COVID-19 has presented the greatest existential threat to our way of life since the second world war. The pandemic severed our normal social connections. In our quarantine solitude, we flocked to the Monster for comfort, fellowship, information, and distraction. Every concern about privacy and the potential dark underbelly of surveillance was swiftly and completely forgotten. In our period of extreme stress, tech became even more central to *every* aspect of our presence on Earth. Tech's intrinsic strengths and flaws became more apparent as governments, institutions, corporations, and people responded (leading to unintended consequences, both good and bad).

It's in that light that we — Paul, Ben, and the many, many colleagues, journalists, academics, clients, colleagues, pals, and family members with whom we interact — have been discussing: "What is happening? What seems to be going wrong? What should we do about it? What kind of world do we want?"

Our exploration is structured around four key pillars:

1. **Capital.** Tech and money are now inextricably interlinked. What has money done to tech, and what is tech doing to money? Is there any way out of a future in which money is the only thing that matters?

2. **Psychology.** We are already cyborgs. Our phones are never out of our hands. Soon, they will be in our glasses and inside our heads. What does this mean? Can we ever disconnect, figuratively and/ or literally? What is happening to our minds (the original, organic CPUs)? Is this exciting and good, or terrifying and a disaster?

3. **Society.** Tech is accelerating the compounding of winners and losers. Live in a zip code full of code? Life has never been better. Live in an analog town? Protectionism and wall building can seem like the only option. Social media echo chambers have started an uncivil war that is currently virtual and digital but could soon feature real bullets.

4. **The New Machine(s).** Artificial intelligence is perhaps one of the most important human inventions. We grant you your cynicism but caution that your raised eyebrows will be your undoing. AI is bigger than anything we've seen, and we've been looking at big things all our working lives. How can it not change every aspect of our world in the coming years and decades? And yet we (the collective we) still have very little grasp of what it is going to do to us.

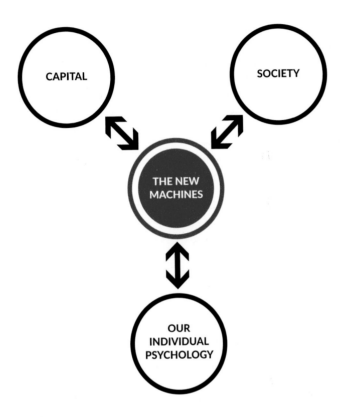

If they weren't entirely obvious before the pandemic, the downsides of our "wired" world are on full display now; a global 24/7 news industry (incentivized to dramatize and inflame every tweet and data point); a social media industry incentivized to amplify the "macro" "breaking news" into "micro" "freaking out"; and the unholy alliance between Mainstream Media and Social Media (Zucker *and* Zuckerberg) that creates a vicious cycle at which Churchill would weep. (Nobody's keeping calm, and nobody's carrying on.) We offer the following in the spirit of honest exploration and genuine humility, and with a sense that without good questions and dialogue, we will never find good answers. That process starts here.

In the tradition of Socrates and Plato, Watson and Crick, Mick and Keith, or Abbott and Costello, we've tried to capture a spirit of our own dialogue, the back-and-forth wrestling with these issues, which are some of the most complicated that humans face today.

In a few pages, we try to net out which questions are important, what we should be paying attention to, and — importantly — what must be done.[2] All of the following thoughts stem from our own research, work with clients, reams of insight, and ideas from others, and hours and hours of discussion and debate between ourselves — on conference calls, in meeting rooms (and a few bars) — where we traded point and counterpoint to develop a narrative that made sense and that could help us find a healthy path ahead.

It should also be noted that in our deliberately short book, we haven't tried to write a complete overview of technology or finance or the human psyche or societal structures. Fantastic but modestly sized books like *The Prince*, *The Art of War*, and *Night* don't completely explain leadership, conflict, or evil. Rather, they each tell a "small" story to illustrate themes of nearly infinite complexity and scope. The story of the Monster is so comprehensive it would take dozens of books to "cover" everything (and would certainly still fail). Instead, we chose to create a small piece to frame one of the biggest stories of our time.

Some will undoubtedly query why we've chosen to focus on "x" and not "y." This is, again, by design. Based on our prior research and work over the past years, we made the best choices we could to focus on what we feel is the *most* relevant and the *most* likely to help us out in the near term. The result is a subjective but informed take on an extremely complex set of interlocking issues.

Our hope is that by allowing ourselves this freedom, we have given ourselves permission to examine and understand the fundamentals of how the new machines are reshaping our economy, our society, even our own minds and, in doing so, get to the essence of what we all need to do to move toward a future that may not be a utopia but is far less likely to be monstrous.

Can we tame the beast?

So if, in fact, we have created a technology monster, how can we begin to tame it?

Throughout the book, you'll find recommendations for policy makers, business leaders, and ordinary responsible members of society. To make things easier (and provide some hope because what follows can get kind of dark), here is a summary of 13 lessons we will *all* need to embrace in the months and years ahead.

1. **Tech is too important to be left in the hands of technologists.** Intimidated by the complexity of technology, "civilians" have tended to let the "nerds" get on with things on their own. How did that turn out? Not so well. Technology needs to be regulated by technologists and a diverse cohort of non-engineers who can't spell "Java." Tech boffins may not like the sound of this, but we'll get used to it.

2. **Real freedom means ending data-based surveillance.** To maintain authentic economic freedom, we have to recognize the rise of covert digital surveillance. It's time to make it much harder — OK, impossible — to extract, mine, and sell our data and information without our true awareness and informed consent. The argument to use sensors, AI, and contact-tracing in the fight against the coronavirus pandemic is powerful, but by doing so, another step will have been taken toward the full compromise of the very notion of privacy.

3. **Treat personal data like your reputation, not your disposable razor.** Centralized data control is the taproot of winner-take-all internet businesses. Some are trying to wrest back control from the digital oligarchs, but that will be difficult. It's up to each of us to

recognize our personal data as a precious asset and *treat it as such* (rather than tossing it onto the web for the gain of others).

4. **Is it safe?** No, it is not. Wondering if you've been hacked? You have.

5. **Get ready for the "splinternet."** Borders used to be drawn on maps. New borders will be drawn according to your IP address. Each region, country, state, or even city (or home) will have different views about the ideal technical, social, and economic model for the future. We need to be ready for different versions of the web, with vastly different conventions for privacy, tax, data, content, access, and more.

6. **Follow — and manage — the money.** Forewarned is forearmed, so simply being aware of how capital, growth, and economic power are changing will put us ahead. Be ready to participate in new regulations, tax policies, controls on capital markets, and more to protect wealth, innovation, and even our happiness.

7. **Stop the brain hacks.** Attempts to use technology + psychology to try to control us and extract value without our consent or knowledge need to be recognized for what they are: a violation of trust and ethics. This practice must be rooted out and shut down with regulation, law, and social convention so our activities, thoughts, and emotions are not hacked and tracked every time we log on.

8. **Kick digital fentanyl.** Tech gives us endless stimulation, a little drip of dopamine with every click, tweet, swipe, and like. It's time to take steps to kick the addiction. Easier said than done, but step one is recognizing the problem.

9. **Just breathe.** Jacking into the web every day, every hour, may not be damaging, but it's not benign. Community, friendship, faith, and mindfulness may seem quaint today, but the path to building healthy identities and minds in the modern age could be paved by actions and practices that have grounded us for millennia.

10. **Learn from the rearview mirror.** Our ancestors struggled to absorb their steam-powered disruption just as we are struggling with our new machines. We should hold up a mirror to the past to see many of the patterns repeating today as tech impacts capital, war, politics, society, labor, ethics, and more. By learning from the past, not

repeating every mistake, we can guard against technology creating dark days ahead.

11. **Be a "rules of the road" co-author.** The velocity and direction of the next phase of the digital economy will be driven by many types of law, policy, and regulation: net neutrality, privacy, patent and IP law, taxation, data protection, industry regulation, AI ethics, labor laws, health data laws, job licensure, sharing economy regulation, etc. Even if it sounds mind-numbingly dull, this *must* happen, and it's up to us to participate and help reset the governance structures of our new machines.

12. **Show agency over your future.** Waiting for "someone else" to figure this out for you — *your* family, *your* company, *your* country — is a mistake. If you're a member of a democratic society, you have the right (obligation, really) to exercise authority over how you manage tech, use tech (don't be a troll), and — critically — how you participate in the democratic process to govern tech. Sitting on the sidelines is for wimps. Don't.

13. **Recognize that "off" may sound enticing, but it's unlikely.** Can we turn off tech? Disconnect? Go dark? Maybe, but it's not easy. For many of us, it's simply not possible. (And what's the fun in that?!) What we can do, *must* do, is reflect on how we want to engage with tech, decide what we want for our societies (and ourselves), and then act accordingly. Maybe we can't turn it "off," but we sure as hell can turn it down!

And here, further, and specifically, we recommend for consideration the following 10 tactics to be instituted without further delay. (Remember, we promised to begin to address the tough, controversial, and important questions!) These are framed within a US context, but they are relevant and applicable in many other countries.

1. **Establish a Federal Technology Administration (FTA).** This organization would sit *supra* to the Federal Trade Commission and the Federal Communications Commission. Akin to the Federal Drug Administration, it would have overall responsibility for creating relevant, contemporary legal frameworks and regulatory licensing for technology for the twenty-first and twenty-second centuries.

2. **Within the FTA, establish a US Data Authority (USDAu).** Akin to how the UK Atomic Energy Authority is a function of the Department for Business, Energy & Industrial Strategy, the USDAu would focus on establishing policies guiding the ethical use of data and algorithms in the commercial sector. It should be staffed with technologists and non-technologists.

3. **Institute data ownership and portability legislation.** Laws should mandate personally identifiable data and meta data to be the property of individuals, not the organizations that capture that data. Individuals must have the right to control their data, including porting it from one service provider to another. They must also have the right to entirely withdraw their data from a service provider.

4. **Institute data and algorithm audit legislation.** The USDAu should have the legal right to inspect the data social media providers have about their users, and how decisions made by algorithms are arrived at. These audits should be made available to the general public.

5. **Prohibit political advertising on social media.** Federal law should prevent any organized group from placing any form of political advertising on any social media platform.

6. **Repeal Section 230 of the Communications Decency Act.** The law should recognize that social media providers are publishers of information and should be held to the same standards as other types of publishers.

7. **Prohibit use of social media by people under the age of 18.** Akin to automobiles, weapons, alcohol, tobacco, gambling, marriage, and military service, social media should have a lower age entry barrier.

8. **Establish Social Media User License (SMUL) legislation.** Similar to driver's education for cars, before being allowed to use social media at age 18, individuals and individuals within organizations should receive training and instruction on its safe operation, leading up to being granted a user's license. This license should be revokable based on subsequent actions and offenses.

9. **Establish federal protection of sovereignty against data incursion.** The FTA would establish federal control mechanisms to intercept and screen out external data entering the US that is illegal, disruptive, and malign.

10. **Overrule anonymity in for-profit social and media platforms.**
    Anonymous speech in the US is rightfully protected by the First
    Amendment to ensure all opinions get a chance to be assessed.
    However, anonymity in for-profit social platforms and media
    channels has led to toxic troll armies, cancel culture, propaganda-
    as-news, and bot farms without accountability. The FTA should
    create a licensing program, akin to Twitter's blue verified badge, for
    individual or organizational contributors to social media platforms.
    Any communications via for-profit social and media platforms must
    be tagged back to the SMUL owner. The FTA should also establish
    and manage a fully open "digital public square" that protects
    unfettered, and anonymous, free speech while filtering unprotected
    hate speech.

This book is a tough-love letter to technology at a time when it has never
been more important. COVID-19 is literally and metaphorically a bug in our
system. In the months and years ahead, we must broaden our minds — and our
policies — to build a better, healthier system, not just to neutralize the bug.
The current (and accelerating) transmission speed of people, viruses, news,
opinion, and capital is unsustainable. Tech is the engine of that transmission.
Brakes are needed before we crash and burn.

We may seem critical, sometimes even harsh, about where we are today, but we
take this tone not because we come to bury tech but to praise it.

We want to save tech, help it mature, help it help us, keep it *On*, not *Off*. Our
path ahead is not to slay the Monster but rather to tame it, to leverage its
power, to reorient technology as a tool for good.

Easy? No. An unfolding path ahead? Yes!

The Fourth Industrial Revolution isn't simply for the folks on top of the Davos
mountains. It's for everyone. This is *our* revolution. This is *our* monster to
tame. *Our* future. Let's make it what we want it to be.

2 MACHINES

In which we look at the past, present, and near-term future of the incredible new machines that threaten to overwhelm us — their creators.[1]

t's easy to point at stories of tech-gone-wild like *Black Mirror*, *The Circle*, and *The Terminator* and say, "Spooky! But that's just fiction." The nonfiction version, though, is more insidious and terrifying. Like a modern version of Shelley's Frankenstein monster, most of the parts of our new Monster are already assembled on the lab table. All we need is the right kind of spark to get it twitching, off the table, and chasing us across town with bad intentions.

These sparks are already flying. Coronavirus and social unrest are adding nitroglycerin.

People with *scopophobia* (the fear of being watched) are going to have a tough couple of decades. If you assume only public cameras and web browsers are tracking your actions and thoughts, we've got some bad news. As we predicted in our first book, *Code Halos* (published in 2014), literally every *thing* is now generating code — data — about you, your parents, your kids, your dog, your car, your home.

Where you go is tracked by the mobile surveillance device in your pocket that we still call a *phone* (which is kind of quaint and adorable). Toys your children play with are spitting out data to help fine-tune targeted kiddo marketing. If your car has a telematics device, your insurance company can know when you're trying to drive after too much wine. One telco leader we know won't use all her company's products simply because she knows how hi-res the picture would be if they connected her browser data, viewing habits, phone history, and smart-home data. Your power company knows when you are cooking, when you open your refrigerator for that late-night calorie binge, how much laundry you do. Clearview AI's software can input a picture of you and connect it with all the other photos of you among a data set of more than three billion other images harvested from social media (most of which *you* shared). Proponents of contact tracing — in the ascendancy as we write — will suggest that use of this technology goes from voluntary to mandatory for the "health of the many." In doing so, surveillance will burst from the shadows and, now more legal and emboldened, strut around the stage on full display.

Virtually every employable skill, the common narrative goes, will be replaced by AI run amok. Our jobs will be automated away, leaving us in a dystopian nightmare scenario where the only human jobs will be oiling and maintaining the Terminators. Driverless cars, doctor-less surgery, chef-less meals, soldier-less armies, teacher-less schools aren't *really* around the corner, but many people are nearly paralyzed about the potential. As we explained in *Machines*, much of this fear is overblown, but the hard truth is, when it comes to robots and jobs, *there will be blood.* . . .

Speaking of blood, the stylized sci-fi of Slaughterbots and Terminators might seem troublingly prescient, but the most likely future of conflict is even more frightening. Hybrid war with smaller groups of special forces operators, pocket nukes, IEDs, lasers, drones, space-based weapons, and rail guns linked with AI technology is warping how power will be exerted. Perhaps more importantly, this smaller scale, easily assembled but highly intelligent and automated weaponry is lowering the barriers to entry to the Big War Show. Nobody needs a particle accelerator or nuclear-powered aircraft carrier to create a massive swarm of unmanned aerial vehicles that link facial recognition software, consumer-grade drones, and small explosive charges. That horsefly buzzing around? It's actually a miniature drone packed with C4 explosives and AI targeting your face. The internet + AI + quantum computing is not online yet, but the prospect is straight up terrifying.

Scared yet? Us too. But how did this happen? It's certain that nobody *planned* to build a technology monster. To understand how we got here, and how we can regain control, we need to go back to the beginning.

Welcome to the web

In 1989, in a small, non-descript office at the European Organization for Nuclear Research, perched on the border of Switzerland and France, a youngish self-proclaimed trainspotter-type nerd wrote a document outlining how the idea of "hypertext" (developed by computer scientist Ted Nelson in 1963) could be overlaid on top of the "Transmission Control Protocol" (developed by networking specialists Vint Cerf and Bob Kahn in 1974) to create something he called the "World Wide Web."

The document, authored by one Tim Berners-Lee, was initially met with head-scratching. His immediate supervisor called it "vague." Within months, though, it began circulating through computer circles like wildfire. It quickly reached academics at the University of Illinois at Urbana-Champaign, entrepreneurs, and politicians in Washington, DC, and venture capitalists in Silicon Valley.

The rest, as historians say, is history.

In 2014, in a list produced by the British Council of 80 moments that shaped the world in the last 80 years, the birth of TBL's baby was ranked #1.[2]

Fast-forward to today, and what *hasn't* the internet touched? Anything, anybody? *We've got nada....*

Of course, the internet, on which the World Wide Web rests, had been developing for decades, but Berners-Lee's insights took what had been a military-academic curio and jammed it right into the middle of everything. And nothing was ever the same again.

The internet/World Wide Web (simply "the internet" in the rest of this manuscript) has become the hub around which every other technology in today's world revolves. Immense global cloud computers, ubiquitous connectivity, supercomputers in our pockets, machines starting to walk and talk, 5G networks rolling out to turbo-charge our connected lives — none of these would have the power they do without the internet at the core. The internet has become the world's central nervous system. And its spine. And its brain.

Obviously, Mr. Berners-Lee can't take all the credit (or the blame). Many, many other brilliant men and women have given birth to parts of the modern tech world and been instrumental in turning science fiction into science fact. Space prohibits too much of that history being revisited here, but there is one person and one idea that deserve special mention in this brief review of how we got where we got — Gordon Moore and his eponymous law.

Moore's Law states that the number of transistors on a computer chip will double about every two years. Simply put, computers continue to get more and more powerful more and more quickly. Never has a law been so aptly named.

Proposed in 1965, when most people had very little idea of what a computer really was, let alone interacted with one, Moore's prediction was a remarkably insightful signpost to the future that set off the arms race we continue to run to this day. In 1965, the IT "business" was a cottage industry that employed a few thousand people in white coats. Today, this business is a global behemoth with nary a white coat in sight.

As a rallying cry and a goal, Moore's Law acted to galvanize and organize computer-based research and development around the world. But nobody really knew — including Moore himself — how long the law would remain relevant. If you'd said 55 years, Moore would have taken that bet. If you'd said indefinitely, even Moore would have raised a quizzical eyebrow.

But here we are today, and although intuitively it feels like our ability to build more powerful machines should start to slow down as we reach the limits of physics (the sheer size of atoms and electrons) and economics (rising production costs as integrated circuits get increasingly packed), new "laws" related to new technologies are coming into effect to keep innovation going (and perhaps accelerating).

One of the most powerful, and potentially chilling, new laws governs the growth of quantum computing.

Entangled in the dark: Quantum computing powers up

On YouTube, there's a short video that shows off IBM's quantum computer.[3] Watching it, you can almost imagine how it felt hearing Marconi's first Morse-code tapping or seeing the first grainy images on a television screen just a few inches wide. After a few seconds of technical detail, we mostly focus on the sound of the gizmo in action. It has the rhythmic, mechanical swoosh-clunk of an MRI machine.

Inside, there is some kind of black art. To exhibit quantum properties, the processor must be chilled to 15 millikelvin, colder than outer space, nearly absolute zero. Frostbit atoms of niobium and silicon become almost completely motionless. And it's dark, shielded so that no light photons or magnetic fields can seep in to mess things up.[4]

And that's when things get really weird.

Unlike traditional computing, based on binary code comprised of zeros and ones, quantum computing introduces a new state — superposition — containing a phase of being both zero and one, which (along with other features like entanglement that even Einstein called "spooky") provides a whole new realm of computational power.

Currently, quantum computers — at their rudimentary level — are mostly impractical; they're sensitive, bulky, require huge amounts of energy, and we're still figuring out the science to make them work. (In fact, a number of skeptical researchers question whether they'll work at all.)

But they *are* real, and their power is increasing. In late 2019, Alphabet (a.k.a., Google) published peer-reviewed research that demonstrated a quantum computer had solved a calculation in 200 seconds that would have taken a classical supercomputer 10,000 years.[5]

From 10,000 years to 200 seconds. Let that sink in.

On this current trajectory, building artificial general intelligence systems, modeling new vaccines and medicines, improving natural language processing, aligning cancer treatments to our personal biology, predicting economic futures, simulating new materials, creating new communications standards, manipulating genetic material (with a turbo-charged CRISPR), developing fully autonomous vehicles, and perhaps even unlocking the secrets of the cosmos are all within reach.

Truly a science fiction world becomes nonfiction with quantum in hand.

Rose's Law — named after Geordie Rose, the founder and former CTO of leading quantum computer manufacturer D-Wave — suggests that the number of available quantum bits (essentially the processing power of a quantum computer) will continue to double every 12 to 18 months. For the past several years, this has proved to be the case. Quantum processing power could be on a path to faster power increases than those observed in Moore's Law.

Rose's Law doesn't contain the time-tested lore of Moore's Law yet, but if it holds true, soon a single quantum computer from IBM, or Google, or D-Wave could outperform all of the existing computers on earth today — *combined*.

Let *that* sink in.

Artificial intelligence has clearly left the laboratory (and the movie lot) and is in our buildings. It's in our homes. It's in our offices. It's pervading all the institutions that drive our global economy. From Alexa to Nest to Siri to Uber to Waze to Salesforce to Azure and more, we are surrounded by smart machines running on incredibly powerful and self-learning software platforms.

And this is just the beginning.

AI power is also advancing at several times the pace observed in Moore's Law. Today's most advanced AI systems are roughly 300,000 times more powerful than those available in 2012, but they need faster and faster "raw iron" to truly come "alive." If we continue on this trajectory, with quantum computing in the mix, we will face systems so powerful that the fastest supercomputers of today would be laughable to the toddlers of generations to come. Raw computational force could fully unleash the long-held promise of AI, perhaps even approach human brain levels of processing power in the forecastable future.[6]

Quantum computing and AI in tandem are perhaps *the* engine of the future, and we stand on the cusp of an adoption curve that will make every previous adoption curve look tortoise-like in comparison.

Speed makes it harder to drive

Looking back at those adoption curves of the past — refrigerators, electric power, telephones — we can see that history does in fact repeat: idea, development, early flat adoption, a steep uptick as the innovation reaches critical mass, and then a flattening of the curve as the innovation reaches saturation.[7] Each significant invention that we take for granted has gone through a similar scenario.

What the following chart so powerfully illustrates, though, is that these adoption cycles are accelerating as successively powerful new technologies are developed. Electricity took almost 50 years to reach mass market adoption. Cars took 60 years to penetrate 80% of the US market. The internet took a little over 15 years to permeate our hearts and minds. Smartphones took less than 10.

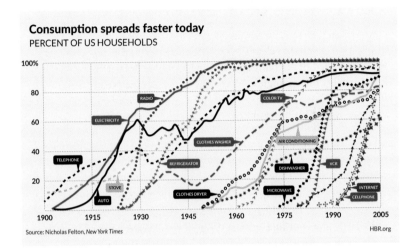

Consumption spreads faster today
PERCENT OF US HOUSEHOLDS

Source: Nicholas Felton, *New York Times*

HBR.org

AI and quantum computing — along with many other advances in memory, processing, software, and connectivity — are set to compress decades to minutes and put incredible new tools in the middle of our futures at a pace humans have never experienced before.

The exponential curves of technology are accelerating every aspect of life so profoundly that the very scale and scope of the change is almost impossible to take in. No wonder we are all struggling to cope with the modern world; no wonder so many people find it so monstrous. Including, seemingly, Mr. Berners-Lee himself.

In 2018, Berners-Lee, who had declined to monetize his document and stepped aside to let two Marcs (Andreessen and Benioff), a Mark (Zuckerberg) — and a whole host of other non-Marc/ks — become billionaires, put his hand back up and said, "The web as I envisaged it, we have not seen it yet."[8]

To Berners-Lee, centralized control of data (through the exclusion of editing functionality in Marc Andreessen's version, Mosaic) created the seeds of a "winner takes all" platform-based internet that, rather than democratizing information and spreading opportunity as he envisioned, led to the dominance of what has come to be known as the FAANG vendors — Facebook, Apple, Amazon, Netflix, and Google.

In 2016, Berners-Lee launched a new company, Solid, that is his attempt to right the ship. Solid is at once a software application and a platform for an ecosystem that provides read-write functionality, permissions, and identity, along with data manageability and real-time updates — the web as originally imagined — providing a platform for a generation of truly empowering applications.

Rather than users' data being owned by the application/service provider (e.g., Facebook), data is stored in a "pod" that is owned by the user and from which usage permission is granted to the provider. In essence, the control model of data usage and monetization is inverted 180 degrees. Berners-Lee's intention is clear — control of the internet by its current controllers is preventing it from becoming the force for good that its proud parent wanted it to be. With further consolidation (and trickle-up economics), and with the imperative to share health and movement data through their platforms, the FAANG vendors' lock on the future will be unbreakable. Perhaps now, we can imagine TBL musing, is the last chance to save his beautiful duckling from turning into an ugly swan.

Whether Solid catches on or not remains to be seen, but that the "father" of the internet has become so concerned about his baby should be a wakeup call the whole world heeds. And this, even before AI and quantum take things to a whole 'nother level. Of course, other alarm bells are also screeching in our ears — though, again, almost so loud we can't hear them.

On the cusp of a "great digital buildout," in which technology becomes embedded in and central to every aspect of modern society, it is increasingly clear that this technology on which the future is being built — with, remember, the internet at its core — is entirely unsafe. The 2016 US election was hacked. The 2020 election is *being* hacked as we write this. North Korea's missile launch program is repeatedly hacked.[9] Film studios are forced to pay ransoms to stop criminals from releasing movies before their official launch. Pacemakers are hacked. Satellites have *(ahem)* probably been hacked. Cars — both the human-driven and autonomous variety — are hacked.

Big Brother was an amateur

Basically, anything with an on-switch can be hacked, and most things already have been. It's no surprise that our new machines — internet-enabled, AI-fueled — are at the center of redefining trust, transparency and privacy. If you have a password to *anything* online these days, you're a user, a product, *and* a target.

If you're wondering if you've been hacked, you probably have been. As Aaron Levie, CEO of Box, put it, "If you want a job in five years, study computer science. If you want a job for life, study computer security."[10] Making our connected homes, buildings, planes, operating theaters, parliaments, bank vaults, classrooms, and virtual reality environments safe and secure is the most important (and limitless) job of the future.

We are all now combatants in the endless war George Orwell warned us about in 1949, only in our case, it's a war of bytes, not bullets. Business leaders, government officials, and each of us as individuals can no longer ignore the new truth that we are each and all responsible for our own security and privacy.

Most of us know some of this, but as individuals, we tell ourselves "I'm just a regular person. I don't have nuclear launch codes, $10 million in the bank, or the formula for Kentucky Fried Chicken. I'm a low-value target for a thief, so I'm sure 'they' will figure it out for us."

Believing in "security through obscurity" might give us comfort, but that conclusion is wrong. Collectively, most of us count on security issues happening to someone else. "That's John's or Li's or Vivek's or Marie's problem, not mine."

But no one and nothing is really safe. The Monster depends on us thinking, "It's all cool" because at an individual level, *we are each right*. Our individual data is essentially worthless, but where our data does have great value is at scale — unimaginable scale. When our individual code is combined with millions and billions of other records, it is a treasure worth more than gold.

The trend of embedding technology into every person, place, and thing had accelerated in recent years, and then the coronavirus came along and turbocharged the trend. A new class and scale of surveillance technology has seemingly gone as viral as the bit of infectious RNA that changed our world. Now we have cameras outside (and *inside*) apartments in China tracking quarantine compliance.[11] Apps are now used for contact tracing to keep us healthy (but also giving away even more personal data for the future). Cameras and sensors are coming online to monitor our temperature, mask wearing, and social distancing.[12] We have remote monitors proctoring our kids as they take online exams.[13]

But the very technology purportedly keeping us safe also leaves us even more exposed. Our wonderful tech — that we (and assumedly you) love — is as sturdy as a shack in a *Wizard of Oz* scale tornado. Fortunes and fates rest on the flimsiest of foundations; even the bluest of blue-chip corporations (and the most deep-pocketed) admit (off the record) that they have been (and continue to be) repeatedly attacked. Our ability to function amidst this truth stems from our individual and collective ability to ignore and deny it.

We used to think of monsters as beasts with fur and sharp teeth in the night. It's hard to think of some patient, smart, tech-savvy young people half-buried in pizza boxes in a basement as being monstrous, but when it comes to security and privacy of *your* data, this is the new normal.

From MAD to MADD

Sometimes it's during a roundtable, sometimes over a glass (or two) of wine, but from Davos to Detroit to Singapore to Sydney and back (via planes or, lately, Zoom), we regularly hear a variation on the question: "What technology *really* scares you?"

Well, imagine a world where *anything* can be hacked. Every pacemaker turned off. Internet-enabled factory machines put into overdrive until they shatter. Every password worth cracking walked through like mist. Air traffic control systems, hospitals, satellites, medical devices, weapons, and power plants all compromised.

Nearly anything can already be hacked, but it's hard work; it takes skill, tools, and the patience of a monk. Quantum computing could blast a sliding screen door into *every* system in less than the blink of an eye. The only defense against malicious quantum computing is another quantum computer. . . .

We're moving now from MAD to MADD — Mutually Assured Destruction (with nukes) to Mutually Assured *Digital* Destruction. In our future, the power of code will be recognized for what it is: as potentially devastating as any bomb or high-tech weaponry. A quantum war would instantly render us a pre-digital society. When quantum computing becomes weaponized, the only truly protected technologies will be waterwheels, hand cranked or horse-drawn. That's what scares us.

This hasn't been overlooked by the major powers of the world. Nobody *really* knows how much investment is pouring in, but China has already publicly committed $10 billion for quantum computing investments, and the country is widely regarded as either on par with or ahead of the rest of the world.[14] The US federal government seems behind in commitment, and only recently announced a $1.2 billion investment.[15] The US National Security Agency budget is classified, but we can assume the organization is not idle. More hundreds of millions (and more likely billions) have been invested in the UK and Europe.

Large banks spend hundreds of millions of dollars on cybersecurity — which sounds good — but in truth, those amounts are typically less than 1% of their entire IT budget (which in turn are typically less than 10% of annual revenues). Given that market capitalizations of hundreds of billions of dollars rest on these foundations, this makes no sense. Unless you're long on pigeons and Faraday cages (or your play is simply to have a high-end crisis management firm on retainer), we all have no choice but to upgrade our cyber defense.

This is the Monster we're dealing with. A hackable, vulnerable, wealth-concentrating, civics-destroying monster growing faster than a beanstalk, set to create many wonderful things but also many terrible things.

What the World Economic Forum hailed in 2016 as the Fourth Industrial Revolution is real and happening now and is creating a time of incredible dislocation — when old ways of production give way to new ones, and when those harnessing the power of the new machine are thriving and those that can't are prone to becoming surplus to requirements.

In the same manner that the First Industrial Revolution was powered by the invention of the loom, the second by the steam engine, and the third by the assembly line, the fourth is being powered by the new machines amongst us — machines unlike any we've ever seen before. The coming digital build-out will be highly promising for the prepared but will steamroll those who wait and watch.

Struggling to manage new technology is nothing new. A little known fact about the original nuclear bomb test in Los Alamos in 1945 is that some scientists believed it was possible that the explosion could ignite the nitrogen in the atmosphere and cause a chain reaction across New Mexico or even the world.[16] AI may not incinerate the world (as James Cameron imagines in the *Terminator* series), but the time has come to grapple with the Monster. We have no choice.

The title of our last book, *What To Do When Machines Do Everything*, may have sounded a bit hyperbolic (clickbait alert!). Clearly machines will never do *everything* — and nobody really wants them to. But it's becoming clearer and clearer that in short order, our new machines will be embedded everywhere and in everything, and will increasingly shape more and more of the work people do today.

The machines we're building are getting more powerful, more disruptive, than any machines made before. It would be naïve to think this is a monster that won't grow any more heads or arms. Disruption is now inevitable. Whether it's positive or negative depends on us.

The new "laws" coming into effect — such as Rose's Law — prove that we have not reached "peak computer" any more than we have reached "peak human." Those who continue to assert that we have already invented all the good stuff are as shortsighted today as they have been throughout history. Ignore them. Fans of Alex Garland's recent Hulu series *Devs* will have gained an insight into the possible futures quantum computing will generate. A "multiverse" awaits. . . .

Technology is no longer the domain of the few but the province of the many. As such, those who win in the next phase of the digital economy are not necessarily those who can create the new machines but those who figure out what to do with them. Ignoring them won't help on the journey.

3 CAPITAL

In which we examine the role money plays in the modern world and the rise of surveillance capitalism — in the West and the East — and ponder the ancient wisdom that love of money is the root of all kinds of evil.

G rowing up, we all hear clichés about the basic rules of money. *The rich get richer, and the poor get poorer. Money doesn't grow on trees. Money makes the world go around. Another day, another dollar. For the love of money is the root of all evil.* These lessons still have merit, but the cauldron of new technology is incinerating the dusty aphorisms of yesteryear.

Money and power have always been inextricably linked. What is new is the value of code, the data we willingly (and unwillingly) share. Since we first traded shells or salt or cows for services and goods, economic value has largely been linear. Work hard, get a dollar; work harder, get two! That was the theory. Now, however, we must deal with the exponential economics of data. Collect 10X the data, *get 10x the money.*

This exponential math knocks industrial-economy economics — based largely on physical products and human services — out of balance. Our new machines are upending the traditional rules of our economy just as the rise of steam-powered industrial equipment shattered the conventions and practices related to water, human, and animal power.

The Monster we've built together allows us to trade our code — yours, mine, ours — for our collective convenience. The data, of course, comes from us. We willingly trade our privacy, inviting ourselves to be surveilled, all in exchange for email, ubiquitous connectivity, cat pictures, "free" communication, personalized content, maps, and more.

Now the companies "too big to fail" aren't banks at all; they are the Code Economy tech giants with massive valuations. As of this writing, Facebook, Amazon, Apple, Netflix, Google, and Microsoft had a combined market cap of $5.3 trillion. For some sense of scale, that's larger than the GDP of entire countries, including Japan, Germany, India, France, the United Kingdom, and . . . you get the idea.

If you think this sounds like someone else's problem — whining and complaining that only matters to the .1% — think again. Unless you're living on creek water and bear meat deep in the Canadian forest, this new set of rules about capital, surveillance data, and power changes *your* life. When the pandemic forced us to hit <pause> on the global economy, the platform leaders got even richer. In the first several months of the COVID-19 crisis, Amazon's CEO Jeff Bezos saw his personal wealth increase to over $200 billion. Five companies — Apple, Amazon, Google, Microsoft, and Facebook — now account for roughly 20% of the total worth of the S&P 500.[1] In the spring of 2020, Apple alone was larger than companies 399–500. These technologies — instrumentation, platforms, social media — gave us comfort, connection, information, and more in a time of crisis, but to imagine that they are designed to function as a public service during a crisis, rather than a commercial growth opportunity, is beyond naïve.

Income inequality changes the operating rules of our economy. Digital oligarchs continue to swell their pockets and change the balance of power by monetizing our private data, and the pandemic was rocket fuel for this rebalancing. Regardless of where you are on the political spectrum, the Monster is altering how our political models — communism, democracy, and everything in between — function, giving rise to nations and corporations empowered not by land and arms but by data. To understand and ultimately confront these dangers, we have to listen to some old advice and *follow the money*.

Income return, growth, and I dream of Gini

Asked to name a scientific formula, most non-scientists would struggle to get beyond $E=mc^2$ or π. Would any of them know r>g we wonder? Unlikely. . . .

Yet r>g is the most powerful formula affecting the lives of billions of people around the world as they go about securing their daily bread and avoiding their quotient of pain. $E=mc^2$ may be famous, but knowing that energy equals mass times the speed of light squared doesn't really help when the monthly mortgage payment is due. Knowing r>g is completely relevant to keeping a roof over your head.

Created by Thomas Piketty, a Parisian professor of economics, r>g explains that the rate of capital return (r) is persistently greater than (>) the rate of economic growth (g).[2] More simply put, capital compounds, and the rich get richer. Confirming what we instinctively knew, Piketty's 2013 best-selling book *Capital in the Twenty-First Century* (Belknap Press) has become the rallying point for an enlightened bourgeoisie that likes making money but knows that the *gilet jaunes* are the sons and daughters of the *sans-culottes* (and fears that Madame Guillotine is waiting in the wings, quietly getting her blade sharpened).

Though r>g was born in Paris, the phenomenon of capital compounding is in evidence around the world. Some would argue that it explains Brexit and President Trump. In fact, it's in America that r>g is really proving most egregious. America was founded and built on the principle that "all men are created equal," yet the streets of San Francisco (amongst many zip codes) see extreme wealth and extreme poverty within feet of each other yet never rubbing shoulders.[3] In the spiritual home of modern technology, equality is a chimera.

The precise role that technology is playing in widening the chasm between a few *haves* and the many *have-lesses* is open to debate. That technology is playing a role, though, is undeniable.

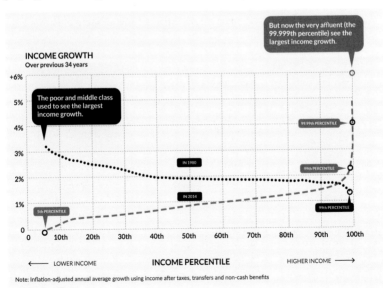

INCOME GROWTH
Over previous 34 years

But now the very affluent (the 99.999th percentile) see the largest income growth.

The poor and middle class used to see the largest income growth.

IN 1980

IN 2014

99.99th PERCENTILE

99th PERCENTILE

99th PERCENTILE

5th PERCENTILE

⟵ LOWER INCOME   INCOME PERCENTILE   HIGHER INCOME ⟶

Note: Inflation-adjusted annual average growth using income after taxes, transfers and non-cash benefits

Source: *New York Times*

If you live in a zip code full of code, life has never been better. If you're lucky and/or good enough to be in the top 10% of earners, your net worth has gone up by almost 200% since 1995.[4] But if you're an average American, in a still largely analog world, your net worth has either stagnated or dropped since the mid-1990s. This may seem great for a few thousand families, but for the +99% not in the <1%, the last 40 years have gotten increasingly difficult, frustrating, painful, and even unhealthy.

Along with Piketty's r>g, another economic formula — the Gini coefficient, named after the Italian statistician Corrado Gini — exposes just how dramatic the concentration of wealth in the US (and other parts of the world) has been in recent years.

Put simply, a coefficient of zero means that everyone has exactly the same income. A coefficient of one means complete income inequality (one person has all the income, and the rest of us get zilch).

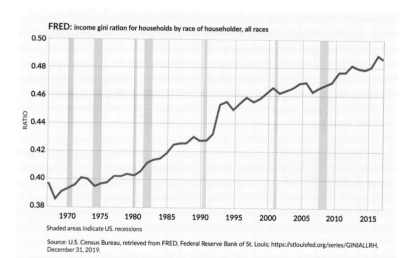

FRED: income gini ration for households by race of householder, all races

Shaded areas indicate US. recessions

Source: U.S. Census Bureau, retrieved from FRED, Federal Reserve Bank of St. Louis; https://stlouisfed.org/series/GINIALLRH, December 31, 2019.

Since the early 1970s, income inequality in the US has grown extraordinarily, as can be seen in the accompanying graphs.[5] The Pew Research Center recently noted that "income inequality in the US is higher than among other advanced economies and has also increased more rapidly in recent decades."[6]

Income inequality is obviously not new, but the outsized impact of technology in the last two decades has been to supercharge the somewhat natural process of compounding interest — what Einstein called the eighth wonder of the world — to an unnatural level.

Economics is called the "dismal science" because it shows up as graphs and formulas, and none of the practitioners can agree on much, but an *economy* is about human beings, and rising income inequality is not good news for the vast majority of we humans.

Lest you think this is just a dry discussion of economic formulas and theories, research by Richard Florida and others shows that higher levels of income equality are correlated with slower economic growth, lower creative capacity, and decreased competitiveness. Moody's also predicted that inequality leads to political turmoil and an unpredictable policy environment (which we're seeing play out daily).[7] Florida's research, in fact, shows that in affluent countries, growing income inequality leads to higher levels of unhappiness for everyone — *even the rich!*[8] Call it jealousy or a rational response to an unfair situation, but when the benefits of growth go to only a few, the collective "we" really don't like it.

Many Americans raised on the legacy dream of the "self-made man" have been loath to look at growing inequality too closely in the face. The founding myth of America was that it was a place to escape the result of compounded capital — kings and queens calling all the shots and collecting all the cash. Yet 400 years after the founding of the Plymouth Colony, compounding has done its job and again produced royalty — Bezos, Gates, Buffett, Page & Brin, Ellison. And now, a new generation of politicians (Roundheads?) is emerging, calling for g<r, and making tech a target in the fight to lower the Gini index.

That tech isn't solely to blame for the woes pointed out by our friendly European economists is clearly true. That tech did a pretty good job of making itself a bogeyman in the 2020 US presidential election cycle is also undoubtedly true. Big Tech has become, at best, lazy and complacent in thinking it was on the right side of history (and the right side of politicians with nice fat campaign contributions) and, at worst, shifty and suspicious by chasing money and influence in some frankly questionable ways. At the heart of this shift has been the surreptitious pursuit of what has come to be known as *surveillance capitalism.*[9]

Surveillance capitalism and the digital oligarchs

In *Code Halos*, we included a chapter entitled, "Don't Be Evil 2.0: Earning — and Keeping — Trust in a Transparent World." In it, we pondered the "dark side" of the emerging data/Internet of Things world, and gave specific advice on how to steer clear of rocks and eddies related to gaps in privacy, ethics, trust, and transparency. Though *Code Halos* sold handsomely, clearly not enough people internalized our recommendations. . . . What we laid out as an unlikely worst-case scenario — that organizations would abuse data privacy and brazenly disregard long-held societal norms of ethical behavior — quickly came to pass as data became the most valuable resource in the world, and accessing and monetizing it became the new game in town.

A *mea culpa*: While we saw the risks, we underestimated the sheer number of bad decisions that collectively would be made. Things have gotten much worse, much faster, than we ever imagined would be the case. The world's response to the COVID-19 pandemic and waves of social unrest now metastasizing across the globe have only served to throw Molotov cocktails on this bonfire of poor choices.

Harvard Business School Professor Shoshana Zuboff has most accurately skewered this new phenomenon in her award-winning book *The Age of Surveillance Capitalism: The Fight for a Human Future at the New Frontier of Power* (PublicAffairs). In it, she paints a vivid dystopian view of unfettered technology molding a society based on weaponized capitalism, powered by the digital "fuel" of our every click, like, swipe, search, buy, tweet, move, key tap, and utterance. She pulls together a remarkable amount of research and insight to paint a Hieronymus Bosch-like picture of how a monstrous economic form, possible only with our new technology, is damaging our society.

Like so much kudzu, the digital economy has grown unconstrained, with surveillance tendrils now digging into every aspect of our personal lives. Our online activities, data we share via health devices, smartphones, geotracking, telematics devices, and even the toys our children play with are all now recording our lives for the monetary gain of others.

One big lesson we've been taught by COVID-19 is that Big Brother was an amateur. Before the pandemic, we traded away a large share of our privacy — via the data we leave in our wake — for access to maps, email, and Instagram. It's a sure bet that we'll trade down even more for access to a vaccine, for information that makes us feel secure, for food, to feel safe and connected.

Zuboff also deflates a common misconception: that if we're not paying for something, *we're* the product. That used to be true, but that was before the rise of the machine economy.

In this economic reality, you and I are not the product. *We* are not actually monetizable. We're largely irrelevant except as producers of code, of data. We're the modern equivalent of prehistoric algae, dinosaurs, and ferns, the buried muck that gets roasted and squeezed deep in the earth's crust until it turns into oil and gas.

It's hard to believe that our browser histories, Alexa questions, Netflix streams, and map usage make up the most valuable asset on the planet right now, but it's true.

Google, Amazon, Apple, Microsoft, Facebook, and others are the most valuable companies in the world because they can snorkel code from every move we make, every breath we take, every conversation we have with a device in listening distance, and convert that data — often without our informed consent — into targeting information for advertisements. "Vote for me!" "Buy this car!" "Puff this vape!" Getting these messages to us drives advertising revenue and a hunger for even more data about what we want and think we need. The more watching and listening that goes on, the more data there is to refine and sell.

Who's buying? Ultimately, we are. Any organization placing an ad for any service, product, philosophy, or political position can — and will — pay for *our* digital detritus to *sell stuff back to us*. Factor in a regulatory environment unable to keep up with the technology, and a digital-economy *laissez-faire* attitude most of us have cultivated, and we end up with our current creepy capitalism. It's not easy — the technology required to pull this off is less than 15 years old — but it's really that simple. Don Corleone and Tony Soprano couldn't have imagined such a racket, but here we are.

Data, privacy, and the health of nations

What may have seemed like esoteric and academic discussions about obscure issues a few months ago are now prime-time questions that directly affect the future ahead of us. How we treat data in our efforts to understand, and ultimately stop, the spread of COVID-19, and how we trade off our privacy in those efforts, will set a course that will be hard to alter.

In China, a sophisticated cell-phone-based "traffic light" system has been rolled out that indicates whether an infected person is clear to move around as normal, or should continue to self-isolate. In Israel, emergency laws have been passed that indicate whether someone may have been physically close to a virus carrier and should thus go into a mandatory 14-day quarantine. In Singapore, Bluetooth functionality on mobile phones is being used to alert people when they are in proximity to someone who has, or has had, coronavirus.

Other governments are understandably stepping up their efforts to replicate policies that seem to be working. In the US, the White House has expressed interest in creating a similar type of health monitoring infrastructure and has called on leading social media platform providers to lend assistance.

Facebook and many tech companies have quickly deployed teams of engineers and data scientists to work on the issue; Apple and Google are building jointly developed contact tracing technology into their rival iOS and Android operating systems. A consortium of leading technology developers — many of the top figures in the worlds of AI and big data — has also formed to apply leading-edge commercial research to the public health fight.[10]

Across the pond, the European Union has called for the creation of a Pan-European app to coordinate a broad response and is similarly shepherding senior tech figures from across a wide range of disciplines. Big Tech has jumped at the chance to restore its reputation and break free from the chains of the tech-lash onslaught it was under.

In the desperate rush to save lives and allow people to again go about their regular lives — and in turn, save economies — data-driven insight and analysis, and social control, are logical and laudable initiatives. Few parents would object if location data from their children's cell phone could help in keeping them safe from harm; few politicians would hesitate to use whatever means necessary to keep those kids safe — *and* keep their parents' votes.

Yet it is precisely at this moment of extreme peril, when our attention is focused on the pandemic confronting us, that we need to think through the even greater peril potentially on the periphery of most people's line of sight.

By opening ourselves up to data-based monitoring, we are further opening the Pandora's box of "surveillance" that has silently crept into our modern world in the last 20 years. Unnoticed by the vast majority, we have entered an era in which governments and technology companies know almost everything about us (based on our online activity), in which privacy has withered away, and which every day inches closer toward George Orwell's vision of *Airstrip One* that consumed him as he succumbed to the devastating influenza virus in 1948.[11]

Health-based monitoring is absolutely the right thing to do in the short-term to help the health of *individual people*, but at what cost to the long-term health of our *society*?

Those who are unaware of the surveillance that surrounds them — or untroubled by it — may ask, "What surveillance?" and, "What damage?" *"We should do everything we can — including phone-based contact tracing — to get the virus under control."* Obscure or philosophical responses about the complicated nature of the trade-offs we need to make between security and privacy in the future melt in the face of the heat generated by finding solutions to the health crisis today.

Those who *are* aware of the surveillance that surrounds them — and *are* troubled by it — are probably destined to be on the losing side of the debate in the next few years. Though the platonic points of the argument can be laid out at length, the need for speed from a jittery public and an impatient news cycle will see politicians from undemocratic and democratic countries alike forgo concerns about the potential misuse of personal data as the fight against the virus is waged.

As such, it is likely that health monitoring will become more and more prevalent in a variety of ways. Cell-phone-based contact tracing will likely become common everywhere within short order. Many countries will perhaps follow Israel's lead and write this into law.

Nothing will trump the widely held view that beating the virus is our top priority, and that if data can help us, then let's use it. It wouldn't be a surprise if soon, something as seemingly untouchable as HIPPA is amended as people anxious about sharing health-related data come to be regarded with suspicion. (This is a twist in the *zeitgeist* anticipated by Dave Eggers' aforementioned novel *The Circle*.)

Most responsible, upstanding members of society will say, "Of course we should use technology to keep us safe." At this moment in time it will be very hard to say we shouldn't.

But with many examples of misuse and abuse of private personal data already in the public domain, with legislative efforts to prop up privacy either nascent or ineffective, and with businesses and governments insatiably hungry for more and more data, surveillance (whether practiced by businesses or governments) is a ticking time bomb that will cause incredible damage when — not if — it goes off.[12]

When health, movement, and biometric data is used to not just see what we look at but also at how we physically, emotionally, and intellectually respond, and when attempts to manipulate those responses become the norm, as they surely will, a threshold into an entirely new world — a world that we were warned about, and which a generation or two ago seemed (to some) implausible — will have been crossed.

We are close to that threshold now. Our response to the virus will take us closer.

To paraphrase Bill Gates's famous phrase, we may be overestimating short-term benefits and underestimating long-term costs.[13] In the world we are entering, we may be healthy, but will we also be free?

Belt, road, and surveillance communism

Though surveillance capitalism is a uniquely American invention — Zuboff places the blame squarely on the doorstep of 1 Hacker Way (Facebook) and 1600 Amphitheatre Parkway (Google) — its refined, perfected form may actually be emerging over 6,000 miles from California in China, home to an explosion of capital and tech the likes of which the world has never seen. Surveillance capitalism is no esoteric techno-libertarian experiment but a playbook for reasserting China's role on the world stage.

For roughly 2,000 years, the Silk Road connected the interior of China to northern India, the Middle East, and all the way to Constantinople and the Roman Empire. That ancient system of dirt paths and ox carts led to China reigning as either the first or second largest economy on Earth from 1 AD until the late 1800s.[14]

Following the Industrial Revolution, Europe and the US have been the world's economic heavyweights. But China wants the title back and intends to do it with concrete and code. The government's Belt and Road Initiative (BRI), launched in 2013, is focused on building the modern-day economy version of the Silk Road by 2049.

China is providing funding, labor, material, and technology to build pipelines, railroads, ports, bridges, and much more in programs across at least 75 countries (including Russia, Pakistan, Singapore, Philippines, Turkey, Hungary, Italy, Kenya, and dozens more).[15] How much China is investing is tough to know, but estimates range anywhere from a measly $0.5 trillion to as much as $26 trillion.[16] Whatever the actual number ends up being, it is safe to assume that BRI is among the largest infrastructure programs in the history of humans.[17]

Although the program's name suggests it's mainly focused on building infrastructure, Belt, Road, *& Code* would be a more accurate name for the initiative. Surveillance capitalism-based technology is a central part of every move China is making, and the country has publicly stated its ambition to become a leader in artificial intelligence by 2030. Many observers would say it already is.[18]

In our book *What To Do When Machines Do Everything*, we wrote:

> Getting [data] is the first step to building effective systems of intelligence. So, first things first; start by ordering chips with everything (well, sensors, actually). . . . These small, often inexpensive devices can be placed on any physical thing and can generate data . . . and then transmit data to a reader. . . . Soon, sensors will be embedded within most if not all physical things in the world.[19]

This is precisely what is happening.

China is now bundling internet connectivity, making huge AI investments, developing space programs, and more, along with more traditional infrastructure projects (including, for example, funding for an AI hub in Malaysia to support innovation and knowledge transfer).[20] To the Chinese, AI leadership and BRI are complementary strategic initiatives. As President Xi Jinping noted in 2017:

> We should pursue innovation-driven development and intensify cooperation in frontier areas such as digital economy, artificial intelligence, nanotechnology and quantum computing, and advance the development of big data, cloud computing and smart cities so as to turn them into a digital Silk Road of the 21st century.[21]

By 2019, the concept of the Digital Silk Road was hardwired into the entire BRI project. In theory, "every piece of infrastructural concrete and steel could be bristling with cameras and embedded sensors linked to remarkably sophisticated AI systems deriving insight and meaning from every activity."[22] Ultimately, as one government advisor sees it, BRI infrastructure gives China a potential "platform for its long-term strategic shift around advanced technologies. This includes electric vehicles, telecommunications, robotics, artificial intelligence, semiconductors, clean energy technology, advanced electrical equipment, rail infrastructure and maritime engineering."[23]

Belt, Road & Code is Zuboff's vision — or nightmare — in its most realized form. China is using new technology to control society. Its leaders are rapidly deploying systems designed to shape social behavior, assign "social credit scores" to citizens, analyze children's faces to see who is paying attention in class, monitor certain ethnic groups, identify protesters, and more.[24] Participation in the social credit score system is currently optional, but in the future it's totally plausible that citizens will be flagged for not sharing their information — again something presaged in *The Circle*. (In fact, some of these social surveillance tactics are now starting to be adopted in the US Mobilewalla aggregates mobile phone data on location, apps used, contacts, device data, and more to study participants in Black Lives Matter protests.[25])

Those charitably disposed toward China's BR&C policy might argue that all this technological infrastructure is being built with good intentions: to benefit humanity.

Maybe, these folks would argue, a connected society that is more rational, observed, and managed — along thousands of miles of Digital Silk Road — will generate more humane, healthy, and productive outcomes than those achieved in the chaotic and unmanageable liberal democracies of the West.

But we doubt it.

By linking politics, business, economics, infrastructure, and surveillance technologies, China has embarked on a mission to create a global system of *surveillance communism* along the Digital Silk Road. Want a port or a railroad or a new airport? Forget the IMF, talk to us! We'll finance your election-winning-jobs-creating-infrastructure-initiative, and will throw in the tools to digitally disarm your opponents, forever! It's a win-win! The Belt, Road & Code Initiative is central to China's vision for the next 100 years — a vision that sets world powers on a collision course.

Tribes and borders in cyberspace

When US President Richard Nixon went to China, little did he (and Secretary of State Henry Kissinger) know that a mere 50 years later, China would be beyond providing the catering and actually eating the US's lunch. And yet, in that relatively short period of time, China has become one of the largest economies on Earth, the fastest-growing economy on Earth, and, with the BRI, the most ambitious economy on Earth.

The internet has brought the Middle Kingdom bang into the middle of any discussion of the future, and if you're not thinking about China policy nowadays, you're not thinking.

So, today, with a shrinking globe awash in fiber optics and cloud computing, what does the not-so-far-away-anymore country of China want?

In the original calculus, it was figured that to keep China out of Communist Russian hands, it was better to have the country inside the capitalist tent rather than outside. Making fridges and auto parts and steel as a subcontractor to Uncle Sam and John Bull would be a win-win for everyone. Western consumers would get cheaper goods. Chinese citizens would create wealth while developing their country, and Western capital owners would get very, very rich. It was hoped that China would remain a grateful junior partner, content to recycle its winnings through the casinos of Wall Street and the mansions of Mayfair, and eventually become a huge new market for Coca-Cola and Bentleys and Siri.

For a while, everything went to script. But then a couple of plot twists occurred.

First, China decided that being the junior partner wasn't so much fun. (This may have been surprising to Nixon and Kissinger, but it wasn't to others familiar with James Clavell's *The Asian Saga*.) It decided that the world of Anglo-American power of the last 200 years was an aberration, not the norm — a wrong that should be righted. So, it started loosening its belt and building roads. It started creating economic infrastructure that mirrored the IMF and the World Bank. And it started a program of expansion (domestically as well as internationally) that was so massive it was almost impossible for the metaphorical sons and daughters of Dick (Nixon) and Hank (Kissinger) to see.

The idea that China wanted to be global top dog was so beyond the Overton Window of the Western global elite, that its ambition was almost patronizingly accepted by those who had developed a bad dose of China Syndrome. That China didn't want to become more like the West, didn't have Western liberal economic desires, was incomprehensible to those who thought that history was over.

At that point, China figured it was better to admit that being *numero uno* was now the real game. Fifty years after Nixon went to China, Xi went to Davos and said, "Countries should . . . refrain from pursuing their own interests at the expense of others."[26]

With China's ambition and real progress in the open, American and European ambition faces its most serious challenge since the 1930s. Can a Sino-American *pax* emerge? Perhaps. Some argue that China would be crazy to do anything that destabilizes its holdings of US treasuries. But recall some sage advice on global politics from Frankie Goes to Hollywood: "When two tribes go to war, one is all that you can score."[27]

The internet and globalization have been mutually reinforcing dynamics since the internet emerged. But now, the internet has pushed China and America into a totally unpredictable position in which literally anything could happen. Even the unthinkable. US treasuries be damned. No more "great humiliation" for us. . . .

All of this — the rise of China, the rise of surveillance capitalism/communism, the tensions between a country keen to expand and a country keen to withdraw (though not cede) global control, with COVID-19 thrown in as *another* plot twist (a black-swan-type one this time) — might be fine, or even better than fine.

But maybe not.

Some will view the scenarios we lay out as merely a rejected script from *Black Mirror* screenwriters. Some may suggest that the Portuguese author Bruno Maçães is a tad melodramatic when he characterizes BRI as "a world of soothsayers, saints, and spooks."[28]

The reality is, nobody *really* knows how this will play out. But we *do* know that surveillance capitalism and surveillance communism are each technical, social, and economic models that are maturing, expanding, and heading toward inevitable conflict.

And in addition, we know that the world is changing profoundly due to the explosion of tech and money and the places from which they stem. We know that old orders and orthodoxies are straining as new centers of power — the ones 6,000 miles apart — replace those that have ruled the roost for as long as most of us can remember.

Power shifts from the G7 to the D7

The countries of the Group of Seven — the leading seven economies of the post-WW11 era — achieved membership in the club through mastery of the leading-edge technologies of the First, Second, and Third Industrial Revolutions. Firmly entrenched in the early days of the Fourth, it's time to ask whether they will maintain this dominance or whether other emerging economies will replace them in the "D7" — the leading "digital" economies of the 21st century.

The short answers are, probably, no, and yes.

Will Italy be in the D7? France? Japan? Or will they be replaced by China? India? The cyber-state of Facebook?

Here's a little thought experiment. Grab a piece of paper and pen. Put your phone down. Now, write a list of apps, technologies, and/or technology companies that you have used in the last 24 hours that are based in Canada, Japan, Italy, France, Australia, and the UK.

How'd that go?

Not well? Got *any*?

RIM went RIP. Olivetti too. Groupe Bull *aussi*. You might have a Japanese TV or a Japanese car but probably not a Japanese phone or a Japanese smartwatch or a cool, useful app, or piece of software.

As a Brit (Ben, not Paul), it pains me to say you probably don't have any British technology around you either. ICL anyone? No, I thought not. (And let's not talk about the last great hope of British tech, Autonomy.)

As we were going to press, Softbank's Arm Holdings was bought by an American chip maker, and the dance-cute video platform TikTok, with its powerful underlying algorithm seen as potentially tipping the board of our global game of *Risk* played with personal data, was being carved up by the American and Chinese administrations.

Now let's list the Chinese and Indian tech companies that millions and millions of people around the world use (or work for): Alibaba, Tencent, Infosys, and Tata, to name a few.

The new machines are changing how wealth and power are created, and it's a shocking and sobering observation that if you created the G7 today on a zero-based budgeting basis (i.e., starting from scratch), over half of the current old guard wouldn't be invited in.

Now the club would be the US, China, and India. Perhaps the UK (thanks Demis Hassabis), Germany (we'll still want nice motors even when they drive themselves, *ja?*), South Korea.

Who else? #7?

North Korea? Russia?

Both somewhat problematic obviously.

Canada? Shopify is on a roll.

Switzerland? Israel? All those crypto valleys?

Sweden? Spotify's pretty cool.

Probably not.

Probably the G7 becomes the D6. Or we let in Facebook.

That the infrastructure of Bretton Woods and the European Union and the Five Eyes is crumbling is testimony to the weakening strength of countries that have fallen behind the technological curve and to the growing power of countries that have used the new machines to take great leaps forward.

Industrial Revolutions I, II, and III went through Manchester, Detroit, and Turin, but IR IV runs through Changzhou, not Paris, and through Bangalore, not Kyoto. And yet, the G7 members — fully paid up and very comfortable, thank you, amid the leather chairs and the silver service — go on as though there may be barbarians at the gate, but Jones or Philippe or Paulo down there on the front desk will be sure to keep them out.

Regardless of how the collision between Big Money and Big Tech plays out, the disruption to economics and power will be vast. Trillions of dollars and zettabytes of data are already starting to flow. Depending on your point of view, this may be great or it might be terrifying, but it's happening largely below the waterline of public awareness.

It's clear we will *not* all agree on any single path. We are hammering the internet into a "splinternet" of different forms, because what is viewed as a monster in one place — a miscarriage of justice, ethics, and power — will be viewed as a friendly giant in another. Regional digital "nations" are forming, and it's likely that this fracturing will continue as sovereign entities apply their own rules about content, commerce, privacy, and politics.[29]

The "Great Relocalization" that was already underway (Exhibits One and Two: MAGA and Brexit) will be supercharged by the threat of viruses — biological and binary — and the failure of supranational bodies such as the World Health Organization to act in a meaningful way, further undermining confidence in the sense of an old-world order. The global pandemic has disrupted the end-to-end supply chain of nearly every good or service comprised by our global economy. Where things come from, how services are provided, how much things cost, every risk equation and more have all become less predictable.

Someday the COVID-19 pandemic will be a bad memory, but as we've been reminded, black swans migrate back around. The timing may be questionable, but there *will* be a COVID-20, a climate change shock, social disruption that overflows sporadic street skirmishes, perhaps a war with both bullets *and* bytes.

Leaders in every company and government have seen how brittle things are, so they are now recalculating supply chain risk for everything from computer chips to potato chips, medicine to mattresses, beer to bog roll. To become more resilient, more anti-fragile, many are deciding to move manufacturing and services closer to home. This won't end globalization, but the brakes are being applied. Technology will be the foundation for a shift toward localization that will reshape cost, taxes, the jobs we have, and the overall balance of power globally.

The only certainty is that the *worst* thing we can do is not pay attention to how technology is driving what will likely be the largest power shift in decades.

*Now, a little
intermission. . . .*

SUNFLOWER: WHEN TECH MEETS CAPITAL

ACT I

A young-ish person sits in a garage/dorm room/
Starbucks and comes up with a cool idea.

Said young-ish person writes some code, finds some
MVP funding, hires a small team, and sets out their
shingle.

Through a completely mysterious and unpredictable
set of circumstances, said young-ish person's
company/app — let's call it Sunflower — starts
getting traction.

Sunflower's funders arrange the next stage of
funding, bringing in other investors — typically
large institutions.

Articles start appearing about the next great
thing in tech — a small, stealthy startup called
Sunflower.

Sunflower's traction goes through the roof.

Said young-ish person appears on the cover of
Wired. Then in articles in the Journal, the Pink
'Un, the style pages of The Old Gray Lady.

The investors invite said young-ish person to a
long weekend of "next-steps brainstorming" at the

Rosewood Bermuda. "Bring a friend," they say;
"Hell, bring a few friends. Or we'll find some for
you."

On the third day of spitballing — after the cruise
on the 100-foot yacht, the dinner with Michael
Douglas, the full treatment at the SENSE Spa — the
investors lay it all out. "You've hit the payload.
You can be the next Zuck, the next Steve. We're
going to make you a gazillionaire."

Gazillionairization begins. The process is simple.
Harvest profile data from the usage of Sunflower
and sell ads against those profiles.

Gazillions flow.

Said young-ish person becomes unimaginably wealthy.
The investors get wealthier yet.

Another great American success story.

ACT II

Q-by-Q growth starts declining from 867% to 789%
to 689%. The investors call said young-ish person.
"Fix it, FAST."

Said young-ish person ideates new customer
retention approaches involving gamification and
behavioral nudges.

The head of data analytics from a Las Vegas casino is hired.

Cass Sunstein and Richard Thaler are retained for a consulting project. AI systems are created to introduce sludge into customer experiences.

The head of sales from a Madison Avenue advertising company is hired. Q-by-Q growth rebounds.

Sunflower surpasses 500 million users. Said young-ish person attends Davos.

Quarterly numbers disappoint. The investors scream at said young-ish person. "Fix it, NOW."

Algorithms are trained to flag and promote inflammatory posts — political, cultural, economic, sexual, violent, fantasy. "Making the extreme mainstream" becomes the whispered joke at Sunflower offsites.

Next-quarter numbers are through the roof. Algorithms learn what drives user attention. Ad rates are increased 33%.

Further gazillions flow.

ACT III

Said young-ish person is still young but not quite
so young. Sunflower is the S in FAANGS.

Said young person spends $7 million a year on
personal protection.

Politicians in Europe and the US demand said young
person answer questions about how Sunflower is
destroying democracy.

Sunflower has more lobbyists in Washington, London,
and Brussels than any other company.

Quarterly numbers disappoint. Investors plead,
"You're spending too much time on externalities."
Said young person, insulated by his gazillions and
his dual-class voting system, stares out the window
and wonders, "When did this all stop being fun?"

Algorithms are tweaked, favoring political ranting.
Rant well, and your post is placed in front of
other ranters. Ad rates for HRS (High Rant Score)
posts are increased 33%.

Further gazillions flow.

Said young person signs the Giving Pledge.

Said young person creates a space exploration
company.

At the next seed, angel, stage 1, 2, and 3 investors' holiday at the Rosewood Bermuda, said young person doesn't show.

Said young person drives cross-country to their parent's house, thinking all the way that Sunflower is about fun and joy and lightness and brightness. It's meant to be beautiful. Something to make your day better, not worse. How did it become a force of darkness and misery? How did I become the scapegoat for all the problems in the world? Why is fake news *my* problem? Why do *I* have to make all the crazy people sane? How can *I* know everyone advertising on the site? Whether they're "real" or not? Honest or not? How can *I* adjudicate arguments between people? Sunflower's a platform — what people do there isn't my problem. Is AT&T responsible for the nonsense people say on a phone call?

HRS posts do even better than anticipated. Ad rates are increased a further 15%. The investors extend their Rosewood vacation and decide to go straight to Davos.

Said person (really not young at all anymore) gives the WEF event a miss, stays home, and posts a picture of their new baby on Sunflower.

And, a picture of the groundbreaking ceremony of the house at Kukio.

<Fin>

 PSYCHOLOGY

In which we look at the consequences of 24x7 tech on our human minds, which are coded to survive and thrive on tiger-filled savannahs and in fire-warmed caves.

N ew tech was *supposed to be fun!* And it was. We connected with old friends, shared pictures of our lives, met new people, debated our opinions, changed our work habits. We even got a little thrill when we thought we were getting something for nothing. (Email! Maps! Chat!) When COVID-19 hit, we all rushed to FaceTime and WhatsApp to connect with family, friends, and colleagues. Zoom went from 10 million to 200 million users per month in the blink of an eye.[1]

But we *aren't* getting something for nothing. We're giving away our privacy and our economic power, and — most insidiously — we're granting access to our minds to people who understand psychology, have incredibly powerful technologies at hand, and want to make *lots* of money from all of those eyeballs. What's wrong with the picture?

If this evokes imagery of brain-eating zombies, you're not far off. There is no solid causal link between tech and our aching heads yet, but there's an awful lot of smoke for no fire. Unhappiness, depression, and even growing suicide rates are increasingly linked to unhealthy technology use.[2]

Consider that from the 1930s through the 1950s, smoking was seen as harmless or even good for us — calming our nerves, aiding digestion, controlling asthma *(really)*. Pseudo-medical advice was often included in advertising campaigns. "More doctors smoke Camels than any other cigarette."[3]

Scientists may have known smoking was deadly, but the rest of us did not (and if we did, we didn't admit it). But truth be told, even awareness of the impact didn't slow us down. A few puffs, and we kept puffing. Damn the pulmonary disease; full speed ahead! Addicted for the long haul. Keith Richards, and others who might know, have said quitting smoking was harder than kicking heroin.[4]

The parallels to modern technology are striking. Try prying the phone from your teenager. Try putting your own device down. Try turning it *off* (more on that later). It's an addiction, the new monkey on our backs — not a diversion. There's a reason we used to call it a "Crackberry."

Mass consumption of new technologies may be the biggest social experiment in human history — an experiment with no oversight — and it's causing real damage. Similar to smoking or other drugs, we've inevitably become addicted to the jolt we get every time we engage the Monster. It's fun until the damage starts. Our *Homo sapiens* minds are simply unable to process — or reject — the ongoing, relentless, toxic stimuli we get from technology all day, every day, from the time we are old enough to swipe right.

Digital fentanyl

The propensity for addiction is as human as laughing or having red hair. It's not known from where it originates or how it is triggered or transferred. All we know is that it exists. At its mildest, addiction might drive sneaker collecting or lottery ticket buying. Higher up — or lower down — the ladder, it drives casino vacations and smoking. At its apex — or nadir — it drives drug addiction. And social media.

Too harsh? Look around you now — what's everyone doing? Staring at their phones. Zooming. On Instagram, Facebook, WeChat, Twitter, TikTok, and a whole bunch of apps your kids know about but you've never heard of. And now we are hooked.

This is no accident. This is by design. The brightest minds of our generation spend 70 hours a week figuring out how to "maximize engagement" and tie users (yes, "users") ever more tightly into the endless scroll — the bottomless pit — of the internet.

"Maximize engagement"? Hook. Addict.

Whole schools of academia exist to train these brightest minds on how to pull off this devil's trick — i.e., make it so we don't even know it exists. The "nudging" or "behavioral science" is now a recognized discipline, staffed by a professorial elite — many of whom are topping up their 401k portfolios with speaking and consulting fees about how to build and run your own digital Hotel California (you know, the one you can check out of but never leave).

This is what the brightest minds of this generation do? Make the brightest minds of the next generation less bright? Wow, the history books of the future are going to be an uplifting, inspiring read. Tom Brokaw is going to have his work cut out for him.

Try putting your phone down. Go on, do it now. Count how long you can go before you can't resist picking it up again.

How long did you manage?

I guess you're addicted too. So are we. . . .

And you're an adult, with willpower and the ability to defer gratification that *your* parents drilled into you.

Imagine what it's like for a 16-year-old kid whose whole life has been a never-ending carousel of instant gratification.

And you're surprised your kid looks washed out in the morning before school? School that now involves even more time staring at a screen.

For parents with teenage kids (that would be the authors), there is a growing, horrifying realization that over the last 10 years, our little angels have been the guinea pigs in a grand game played by corporations "maximizing engagement" to maximize money, with little or no regard to the consequences.

We parents were so in love with cool tech ourselves, we thought it hip and helpful and safe to get Johnny and Jane Angel a phone, with a similar disregard for the unintended consequences. The first little emoji text we got from them was so sweet. We didn't realize that first text was going to turn into 100, then 500, then 1,000 — an hour.

Forgive us, Lord, for we know not what we do.

Literally, we do not know what we have done.

This didn't happen according to some premeditated plan. Nobody set out with the intention of creating a technology that hooked us, depressed us, and shattered our personal identities. The inventors weren't evil; they were trying to create something cool, make something their friends loved and used. And it worked!

But in just a few years, the systems have grown so vast and so powerful that we're now certainly past the point where we can continue to justify *naïveté* and youthful exuberance. It's time to admit that the inventors, company leaders, and consumers — yes, us too — of these new technologies all know what we are doing. We are *complicit* in allowing these technologies access to our minds.

When the pandemic hit, we all reached out for each other, a purely human reaction. But we had to do it via Facebook, Twitter, Netflix, Zoom, and WhatsApp for comfort and support, for protection against isolation and loneliness and fear. Nearly overnight, usage of these platforms doubled and tripled. Humans have had war, plague, and trauma throughout history, but we have never used technology to link to each other so intimately and at such scale.

We got the safety and community we needed, and that's great. Who can even imagine how much worse sheltering in place and quarantine would have been without technology lighting our days and nights, helping us continue to work, letting us connect to each other to share our fears? In just a matter of weeks we simultaneously became more separated — physically — and connected — digitally — than ever before in history.

Now we're waking up, but the genie is not just out of the bottle; it's already wrapped itself around the world like a, well, a worldwide web. Or maybe, more poetically and metaphorically (and heart-breakingly), it's just standing there on the corner of your street. On the corner of every street. Like a drug dealer. Knowing the users will be along in a second. . . .

Because they're antsy. Addicted. Hooked.

We are undoubtedly more linked because of the pandemic, but we are now also more exposed and dependent on the Monster. Nothing but screens for months made it that much more difficult to withdraw, to connect less via tech, to kick.

This is great news for the platform powerhouses, but may be a problem for the rest of us. The economic spoils of being able to get inside our heads, to nudge our behaviors related to buying, voting, or building social movements, are now even harder to turn off and large enough to reshape our global power structures and economy.

But it gets worse. We're not just being manipulated. The way we use technology now is actually *harming our minds*.

Your 70,000-year-old operating system is melting

The operating systems we use in our daily lives — Windows, Android, iOS — are regularly upgraded. But our own personal operating system, the "code base" of our minds, built and tuned over millions of years, hasn't gotten a major upgrade since the "cognitive revolution" 70,000 years ago.[5]

Originally our "software," improved by millennia of natural selection, helped us avoid being eaten, cooperate with others, invent tools, make babies, and ensure the survival of the fittest hominids.

And it worked! Our brains — the most complex mechanisms in the known universe — succeeded spectacularly. We discovered fire. We built the Taj Mahal. We learned to carve, and paint, and make music. We got Mozart and John Coltrane. But now in the modern world, our minds are running on legacy software that is being systematically hacked by the very technologies our brains created.

Ungrateful HAL. Ungrateful Siri.

Important subroutines that protected our hominid ancestors as they went scruffling across the savannahs are still running today, and, as they collide with technology more powerful than our aching heads, this is becoming a problem.

The landmark book *Thinking, Fast and Slow*, by Nobel laureate Daniel Kahneman, paints a picture of just how miraculous our minds really are, but it also shows just how easy they are to deceive or "nudge" by dialing into our biases — how, in effect, "programmable" we are.[6]

- **Optimism bias** helps us believe in the future, but it drives horrible decisions. In a study of entrepreneurs, 81% believed they would succeed (and 33% said they have 0% chance of failure), even though the five-year survival rate of small businesses is around 35%. Optimistic rock star CEOs actually underperform in valuation and operating performance because they don't assess risk well.

- **We are hardwired to seek out bad news.** We know now that so-called "negative" emotions, such as fear and disgust, and the threat of negative consequences such as losing a golf game, making failed investments, or failing loved ones, are stronger motivators toward action than the possibility of achieving gains. Our actions are driven by brains programmed for negativity dominance, so we give priority to bad news.

- **We aren't good with probabilities.** Humans consistently and predictably overweight small-probability events. It's why we pay so much for life insurance and consistently hang onto a "sliver of hope" even when the probability of a positive outcome (say, in a lopsided sporting event, serious illness, or contested election) could be exceedingly low.

As early Facebook investor Roger McNamee noted, "The incentives to manipulate attention are all about preying on the weakest elements of human psychology."[7] Now, many are exploiting these insights and our new technologies to gain power and economic benefit via large-scale *brain hacks*.

Facebook runs "A/B tests" on users, often without our knowledge, that link psychology and technology to keep us clicking and swiping like junkies. Innovations like the "infinite scroll," which makes an endless stream of content available without any friction, are not an accident. These are highly sophisticated technologies designed to keep users locked in, engaged, even addicted (and this is *according to the people who invented them*).[8]

At first, these innovations were used to serve us content about our family and friends that seemed harmless enough. Then they were used to sell us soap powder, which we sort of tolerated, inoculated as we all are to modern advertising. Then they were used to sell us politicians and political causes, via ads often dressed up as "news," micro-targeted to those of us identified as "persuadables."[9]

And that was when we all were introduced to Cambridge Analytica.

Funded by Big-Money political puppet masters, Cambridge Analytica used data harvested from a seemingly innocuous quiz on Facebook that many thousands of people voluntarily took (because it seemed harmless enough) to build profiles on millions of people who were then fed reams of politically oriented advertising that, some have argued, played an instrumental role in the 2016 US presidential election and the Brexit vote in the UK.[10]

Whatever the truth and consequence of what actually happened, what is indisputable is that the Cambridge Analytica moment brought into relief just how powerful social media technology is and just how nudge-able the average 70,000-year-old operating system really is.

Anyone with a teenager has a front-row seat to how the process unfolds, demonstrating the psychological impact of loss aversion on self-esteem (she'll know it as FOMO), as they are simply unable to stop texting or seeking that elusive #InstaGood selfie. Adults aren't immune either as we seek that zap of adrenaline from the syringe of fickle Facebook fame. Psychological insight plus micro-targeting from platform companies can debatably turn elections, but it undeniably keeps our credit cards melting as we now spend nearly $3 trillion a year online.

Because anxiety paid off for our early ancestors, it's hardwired into us, as well. Paranoia dealt prehistoric humans a better genetic hand because they didn't get eaten, sick, or hurt as much, leading to more hominid babies than their more easygoing neighbors. Today, if our minds don't have to worry about us being eaten, they still search for problems to be anxious about. The internet provides! (And bang into the middle of this stepped COVID-19, Brexit, and social protests driving us even more toward the warm glow of our screens.)

Humans and other animals watch each other to see if we're missing out on better shelter, more food, or a more desirable mate. If we decide we're less well-off, we get bummed because we are hardwired to see ourselves relative to others.

This has helped us survive as a species, but now we are simply cooked *sous vide* in daily app-based reminders of our shortcomings — because nobody can *really* keep up with the Kardashians, Cameron Dallas, Selena Gomez, and PewDiePie.[11] And nobody can ignore the drip, drip, drip of terrifying news about elections, the pandemic, and climate change swirling in our anxious minds from the second we wake to the moment we fall sleep (and then in the dreams we have in the night).

Our screens are the windows into the mechanical clockworks of the Monster. The act of engaging content through the screen is problematic because our minds tell us the image we see in our device's Gorilla Glass screen is as real as the world outside our windows — the *original* Gorilla Glass.

The QAnon conspiracy or propaganda or deep fake or endless terrible news and catastrophes on our Twitter feeds seem every bit as immediate and authentic to our minds as the *actual* birds tweeting in our yard or the bustling city that we can see through our window panes. We believe the screen world is equivalent to the window world — except it's not. That's perhaps the biggest mass cognitive distortion in history. The screen world is only a curated approximation, a perversion, a fabricated golem substituting for our complete, messy, nuanced, beautiful real world. Our screen world is more like a fever dream — incomplete, ephemeral, less than real, uncontrollable — than our actual waking lives. To see it as anything other than that lessens our engagement with what is real.

The internet and related technologies promised us connection, community, and coordination, but Herr Freud (from our freshman psych class) has collided with Gordon Moore, whom we met earlier. What we've built has left us with depression, election tipping, trolls, and casual cruelty.

So who's happy now? Nobody really. Economist Marco Annunziata was among the first to raise the alarm, noting that "our cognition itself is under attack" by toxic psycho-technical systems being deployed for economic and political gains, leaving us in the grip of "the great cognitive depression."

> When trillions of things not only collect billions of bits of information but also demand our attention and change our environments dynamically on the fly, our ability to think, make decisions and take actions may be on the verge of collapse.[12]

These trillions of things and billions of bits demanding our attention are in every internet-connected device increasingly found in our cars, offices, homes, and clothes. Soon, our glasses will harvest data from us and feed us all the news fit to look at. Soon, meshes and implants will bypass our five senses and serve information to us through direct shots into our cortex. But before we get to that, we have the *primo* invasive technology of all time right in front of us, right now. In fact, you're probably reading this book on it.

We speak, of course, of our phones. From their introduction through 2018, roughly 19 billion smartphones have been sold.[13] Many of us are already on our fourth or fifth iPhone or Android device. As we saw earlier, it took cars and fridges and radios and TVs decades to reach mass-market adoption, yet the smartphone did it in the seeming blink of an eye. In 2016, *Time* called the iPhone the most influential gadget of all time.[14] It would be hard to disagree.

Smartphones have become so successful because they're really useful, cool, and good. Basically, they've become our bank, record collection, books, maps, virus-tracking system, letters, TV, radio, cinema, concierge, doctor, travel agent, keys, status symbol, calendar, etc. . . . oh, and our phone.

But paramount among these factors is the phone's collision with social media. At the time of the iPhone launch in 2007, Facebook had 20 million users — an incredible achievement no doubt, celebrated by Aaron Sorkin, among others, but still a minor element of the broader internet story.

Fast-forward to late 2020, and Facebook has 2.7 billion users.

Smartphones and social media exist in a double helix that has become part of the DNA of *Homo sapiens*. Now a phone with Facebook or Twitter or Instagram or Snap or TikTok is a $25 \times 8 \times 366$ addiction machine that none of us — not POTUS, not you — can put down.

At the heart of all addictions is a secret story, tough to face, tough to tell. It's easier to cover it up with the narcotic of choice — for many, a snarky tweet, an Instagram picture of dinner.

Social media has unlocked the secret story of mankind's narcissistic rage that people have felt since time immemorial but that historically has been repressed (into back pain or atrial fibrillation) or expressed in other ways (abuse and cruelty at a micro scale, war at a macro scale). It's a narcissistic rage that people feel toward a world indifferent to them and that doesn't recognize their talent/genius/uniqueness.

Now this deeply human characteristic can be funneled straight into the phone at the moment it's felt. Social media — ever present in your hand in the socially distanced checkout line or the car— has become the medium for all the world's rage: unfiltered, unfettered, uncontrolled by the traditional inhibitor of having to say something to someone's face. Now, Instagram monetizes the 16-year-old girl's rage that she's not Kim Kardashian. Twitter monetizes the rage that the 50-year-old man feels that he's not rich enough and that the "elites" are more interested in the rights of transgender folks than him.

The inventors and operators of social media and smartphones, of course, had no clue what demons they would unleash. They didn't know what their inventions would run into — but they have run into them. And through the whispering voices of the Vampire Squids, they have monetized this up the wazoo.[15] Smartphones and social media have become "all the rage" through unleashing "all the rage" of the world.

Anger, surveillance, and premeditated cognitive overload are not the only by-products of our little rage machines. The Monster is actually reshaping how we craft our own identities, how we construct our *selves*.

Who are you again? Identity in the digital age

Although we may not like to admit it, our identities have *always* been sculpted by the people and messages that surround us — our families, congregations, nations, gangs, communities, schools, kingdoms, the music we hear, what we read, what we watch, the news we roast in each day.

For as long as we've had the written word, media has played a significant role in shaping our identities, but now technology is fundamentally changing how we craft the individual self-portraits we each share with the world.

Technology now gives us immediate connection to any group, any tribe, any club, any belief, anywhere in the world. We have unprecedented ability to inform, configure, share, test, reflect, amplify, and adapt our own identities with a velocity like never before. We can compose who we are, be who we want to be.

This is fantastic. When we are feeling safe, fulfilled, and energized, we can share these feelings and amplify them. We can access our chosen tribes no matter how big or small, no matter how near or far. When we are lonely, alienated, or scared, we no longer have to wait until morning for comfort and assurance from our families, our schoolmates, our congregation. The web can help us suffer together through teen *angst* (and middle-age *angst*, and elder *angst*). The opinions and communities we crave are always there, just a swipe away, without constraint from those who may oppose our dress, opinions, abilities, race, whom we love, how we pray.

On the other hand (and you knew this was coming), there's the dark side — the rage. The antiheroes of our stories and holy texts — Satan, Loki, Anansi — are tricky. Tech is the same. "Be whoever you want!" is the ultimate freedom, and the ultimate seduction. Technology gives us access to something powerful — compassion, community, collaboration — that helps us build identity in ways that we couldn't before. But these gifts come attached to a funhouse mirror that, if we're not careful, can shrink what is good and magnify our darker natures.

> **"You be you!"** *<no matter who gets hurt>*
>
> **"Everyone is doing it!"** *<even though it may not be healthy>*
>
> **"YOLO!"** *<even though you may be cutting your life short>*

Racists, conspiracy buffs, incels, enraged trolls, antifa extremists, misogynists, anyone aching to belong to something — anything — that makes our pain and isolation fade for a moment can now easily find others to agree with us, fill the burning holes in our souls, soothe our minds. The vast, anonymous web can connect us to others who agree with us, but it also leaves us without the guardrails and guidelines historically provided by the family, tribe, religion, books, teachers, and friends that have traditionally helped us learn the "right" path.

Early evidence shows that this unfiltered and unfettered rage is having serious unintended consequences. Nearly a billion people globally now suffer from depression or substance abuse.[16] "Deaths of despair" in the US — suicide, alcohol-related, drug abuse — have been on the rise for decades but have rocketed up since 2000.[17] Depression is on the rise all over the world. A growing body of evidence shows that the number of hate crimes continues to rise along with social media use. The pandemic-related lockdown has driven us further into the arms of our technology and further into the depths of this rage.

Is technology fully to blame for this carnage of identity? Of course not. Is tech *really* changing how our minds work, how we see ourselves, how we assess our self-worth? Of course it is. So did books and television, but never before in history have we each had so much power at our fingertips to author, edit, and publish and re-publish our own identity . . . 24 hours a day, every day.

Perhaps Keith and Mick put it best with their searing rock-and-roll prayer seeking protection:

> The flood is threat'ning
> My very life today
> Gimme, gimme shelter
> Or I'm gonna fade away[18]

Technology gives us the means to find *both* shelter *and* flood, but the choice — as always — is ours. Thousands of years ago, the ancients wrote about our struggle to create a virtuous character and identity. This struggle is part of the human condition, but tech — with its instant validation of *any* idea — has made our path even more difficult to navigate.

What to do? Can we fight back, find shelter, not fade away? In the pre-digital world, our identities were molded by faith, family, meditation, philosophy, friendships, familiar rituals, affiliations. Discounting these sources of healthy identity has been a mistake.

"But wait!" you cry. "I'm a modern person. Meditating with my eyes closed, reading 2,300-year-old class notes written by Aristotle, joining a book club, genuflecting to an invisible sky-dweller — that's fine for others, but I don't have time for that crap!"

You be you, sure, but look at the data. The path we're on is not healthy; it's toxic.

We're honestly *not* going all Newest Age post-techno-hippie on you. We're *not* saying, "Just find the off-switch and assume the lotus position." We know we can't completely pray, hug, or meditate away the dark forces of the Monster shaping our identities. But it's time to face up to the impact technology is having on our inner lives, our minds, who we are.

Jacking into the web every day, every hour, may not be damaging to our identities, but it's not benign. Community, friendship, faith, and mindfulness may seem as nostalgic as a Norman Rockwell painting, but they couldn't be more important in the modern age. The path to building healthy identities and minds in the modern age is increasingly paved by actions and practices that have grounded us for millennia.

The singularity is near. Unfortunately

Descartes and his *frères* invented the equivalent of an endless metaphysical drinking game about whether our minds and bodies connect or if they are independent from each other. Philosophers, theologians, and scientists have played this game for centuries. The newest technologies of the Monster have made it clear to us that the linkage between *wetware* (the unsavory name tech engineers sometimes give human users and our brains) and *hardware* is rapidly becoming inextricable.

Breathless predictions of mankind's merger with machines are of course legion and stretch back into antiquity. As noted, Shelley's monster brought — with a jolt of electricity — an ancient idea into the popular consciousness during the First Industrial Revolution. More recently, Ray Kurzweil's concept of the singularity — when technology becomes super-intelligent beyond human abilities or even our imaginations — has placed the dream/nightmare of transcending biology squarely in the middle of the Fourth.

Elon Musk's Neuralink — launched quietly in the summer of 2016 — has quickly made progress in attempting to turn science fiction into science fact. Fifteen years after Kurzweil suggested the singularity was near, he may finally be right.

The Neuralink — a tiny computer stitched into the brain via fiber-optic, electrode-rich wires — can decipher neurological signals and translate them into physical motor skills. It can also connect with external Bluetooth networks, so instructions can be bi-directional; the human can "think" what it wants an external machine to do, and an external machine can direct the human to undertake what it wants it to do.

At this point, we're sure your first thought is exactly the same as ours: What could possibly go wrong?

Beyond any concerns we'd have about implanting an experimental, unproven device into our bodies, or fears of rejection and infection, or anxieties about needing an operation every time the system goes wonky and needs fixing, the simple notion that this giant leap for humankind makes us merely a node on a network — susceptible to all the issues of cyber insecurity that we've examined through this book — is truly terrifying. For as we've shown, every positive breakthrough in computer technology in the modern era has also brought with it a dark side, in which human folly and sin has reveled with the possibility for chaos and mayhem and money.

Kurzweil and Musk and their fellow utopians all acknowledge the possibilities of misuse and abuse of their ideas and innovations, but they brush aside concerns by playing the Luddite card. Any objection to technology progress is the insane, retrograde rambling of a lunatic. The mad professors nowadays are not the mad professors but the sane ones in our Alice in Wonderland upside-down world. Mad is bad (good), and sane is lame (bad).

Bear in mind the last great biotech leap forward — Theranos — and how that all ended. Elizabeth Holmes dropped out of Stanford at 19 to create a startup that promised to revolutionize and disrupt the slow-moving and pre-digital market for blood diagnosis. Billions of investment dollars flowed her way, and tier-one alpha movers and shakers (no names — a roll call would be piling on) lined up for photo-ops alongside Holmes. In 2015, Theranos was valued at $9 billion. A mere three years later, a federal grand jury indicted Holmes on accusations of falsifying results, and the gig was up.

The tale of Theranos is not one of *technical* hacking but of *principle* hacking — that moving fast and breaking things and wrapping things in a reality distortion field and faking it until you've made it are all legitimate and just the way things are done on the leading edge of innovation and progress. That is what is so disturbing about the notion of the man-machine interface — that the culture of tech, with all its shortcuts and pivots and v2s will overwhelm the "do no harm" culture of medicine, and "mad is bad and sane is lame" will become an operating principle in a world of real operations. Hacking a server or an iPhone or a nuclear reactor is one thing — hacking a person is another thing entirely.

Of course, as cool, groovy futurists, there is a part of us — and no doubt you — that thinks the Neuralink sounds cool. (Of course, we may have already been hacked and programmed to think that.) But, as we keep demonstrating throughout this book, cool, groovy ideas keep getting less and less cool and groovy. If insanity is really doing the same thing again and expecting a different result, we can't keep arguing that new tech should be given the benefit of the doubt when time and time again, we see innovation after innovation go bad.

Blood-hacking or human-hacking or biohacking or whatever the singularity eventually becomes is an idea whose time should still not yet come. It's something that should be put back on the shelf in the R&D lab until we can ensure hacking is not the inevitable fate of all new technology. A hacked human is a fact that should remain fiction.

5 SOCIETY

In which we explore the fraying of civilization as technology turns from meaningful miracle to malignant master, examine the possibilities for cyber wars to spark "real" ones, and look to history for answers to how the future will unfold.

We're clearly in the early period of the Fourth Industrial Revolution. Society — and our overall geopolitical environment — is already being reshaped even more than we may recognize by the compounded impact of new machines on our capital and individual psychology. Already, a groundswell of modern Luddites is lashing out at the power of new technology. We're also seeing a new kind of war — fought with code and bytes rather than bullets and bombs — right in front of us yet beneath the horizon of our awareness.

One of the biggest questions we face today, and perhaps the most concerning, is how our new machines will change the topography of power across the globe. Throughout human history, societies have been birthed, shaped, and exterminated by the use of hard power. It's a cold realization that previous seismic shifts involving capital and technology have often led to a "hot" war. The First Industrial Revolution allowed us to use steam to make all sorts of other machines, contributing to the American Civil War (among other conflicts). The North was more highly industrialized — with people using machines — while the South was still almost exclusively agrarian, powered by the labor of enslaved millions. The question of who got to legislate enslavement in the new territories was one of the causes of the ensuing conflagration. The Second Industrial Revolution helped fuel the horrors of mechanized combat in World Wars I and II.

Our challenge, our grand opportunity, is to navigate our way forward toward *La Belle Époque 2.0* while avoiding modern versions of Scylla and Charybdis (the pair of maritime monsters who antagonized Odysseus on his ancient journey home).

There's no question our modern society is under threat, but all is not lost. Even a cursory look at recent history — and we think 1848 is a pretty good analogous pivot point — gives us a preview of coming attractions. In the past, things got horrible, but then they got better (although admittedly not for everyone). We have a chance to follow that path toward the light, but we would do well to take lessons from those who came before us.

Modern Luddites and the growing techlash

Anyone with a pulse and a feeling for the *zeitgeist* can tell that the tide is turning. Has turned. Until recently, new technology may have been seen as mostly benign — music, TikTok dances, car rides, Zoom calls — but as its power has grown, its perceived threat has increased. Now, people, organizations, and policy makers are rapidly organizing against our modern machines. With or without COVID-19.

- **Autonomous cars take a beating.** In Arizona, people have slashed the tires on autonomous vehicles and, in one case, even threatened a driver with a gun in reaction to a fatal accident involving a "driverless" car.[1]

- **Google search finds rocks.** Prior to the work-from-home mandates, San Francisco residents frustrated with the impact Google is having on local communities (and income inequality) attacked buses full of Googlers going to and from work.[2]

- **Taxi drivers pump the brakes.** In the pre-COVID days, taxi drivers protested Uber by blocking traffic in London, Paris, Jakarta, and many other cities. In South Korea, several taxi drivers tragically set themselves on fire in protest of ride-sharing apps.[3]

- **I'm ready for my closeup.** In late 2020, arch-leverager of user-generated data Netflix developed a hit film, *The Social Dilemma*, exposing the horrors of leveraging user-generated data. The irony was lost on most commentators. . . .

- **Break 'em up!** In a highly polarized political environment, it's noteworthy when both sides of the aisle agree about anything. Politicians and citizens from the left and the right — in Europe, India, the US, and other countries — are increasingly open to considering regulating, and even breaking up, the massive tech companies that some see as "too big to fail."

These incidents, and countless others, are symptoms of a groundswell of rage against the new machines.

If this sounds familiar, it should. In the late 1700s, Ned Ludd and his compatriots famously destroyed knitting machines powered by waterwheels. Ned and his colleagues, known as Luddites, were afraid that mechanical automation would disrupt common human labor practices and ultimately cost them their jobs.

We hold the Luddites up as good common folk railing righteously against the dark forces of capitalism and evil machines. Sadly, in the end, the uprising was put down by force.

What often gets overlooked, however, is the *other* sad fact that the Luddites were actually right. They *did* lose their jobs and livelihoods. The machines were too powerful, too cheap, too effective, and could not take up arms against senior management.

Through the lens of history, the Industrial Revolution provided the machines and productivity improvements to help society overall, but it was certainly not all roses (see: child labor, pollution, tenements, unsafe working conditions). At the time, people who saw their productive jobs get automated away were not thinking about how much better off their great-great-great-granddaughters would have it. They were focused on putting food on the table tonight, just as people are today. A *gilet jaunes* protester in modern Paris put it best: "Some people can afford to think about the end of the 'system,' but most of us just worry about how to cope until the end of the month."[4]

The sound of wooden looms being splintered and the shouts of rebellion against automation echo today through recent punch-ups against technology, but history offers some clues as to how this will all play out.

Want a hint? The house money *never* bets against new technology in the long run.

In *no* case have we collectively decided to simply turn the machines off. In *every* case we have collectively chosen to move ahead with the new gadgets. That's what humans do. We invent, we break stuff, and we always lean toward the new while trying to manage the inevitable disruption.

But we also try to manage disruption so as to make it less disruptive. Regardless of which side — capital vs. labor, the future vs. the past — we may be on, we shouldn't see our modern Luddites in an entirely negative light. Arguing about privacy laws, IP protections, taxes, and even throwing rocks at the Google bus, are a kind of public discourse around how we will manage these machines. People on the front lines of disruption, many of whom may not have many other options, are actually doing us a service by forcing a decision on who we want to be as a society.

Facebook and others are, to their credit, now actively supporting efforts to better regulate social media platforms. That's laudable, but tech business leaders of publicly traded companies are *required* to maximize the value of their firms. It's a timeless *cliché* that "turkeys won't vote for Christmas."

The *derecho* of bad news that buffets us every day may feel overwhelming, but in some ways, it can also be seen as an optimistic harbinger of admittedly disruptive, but potentially positive progress driven by technology. The specifics of how this techlash will play out remain to be seen, but it is real, and it will certainly grow in magnitude, intensity, and duration.

In this, perhaps the patterns of history are our greatest teacher.

Lessons from the rearview mirror

The year 1848, give or take a few years, was the end of the beginning of the First Industrial Revolution. Think of the early years of the revolution as the earthquake. There were certainly disruptions as new machines — looms, railroads, etc. — were invented, but the tsunami-sized impacts to how our forebears lived their daily lives hadn't yet begun to reach the shore.

Social upheaval, political shifts, new ideas about economics, law, immigration patterns, monetary policy — nearly everything was up for grabs across the

Aspect of daily life	1848 (give or take a few years)	Roughly today
Socio-political shifts	• The Spring of Nations occurs, encompassing more than 50 uprisings across Europe, including in France, the German states, the Austrian Empire, the Italian states, Denmark, Poland, and many others. • Women's rights movement ignited by Elizabeth Cady Stanton's Declaration of Sentiments. • Beginning of China's "Century of Humiliation" in 1839.	• The rise of economic nationalism rises globally (Hungary, Italy, US, Brazil, UK, Philippines, etc.). • Brexit. • China initiates Belt and Road Initiative. • Arab Spring (not ending in a democratic Utopia). • The COVID-19 pandemic (and subsequent economic shock). • Social change movements (e.g., Black Lives Matter globally, anti-authoritarian protests in Hungary, pro-democracy protests in Hong Kong, anti-corruption protests in Lebanon, and more).
Technologies	• Wide dissemination of Industrial Revolution machines by the early to mid-1840s.	• Cloud computing, social platforms, AI everywhere, big data, AR/VR, quantum computing (coming soon!), contact tracing apps.
Economic ideas	• Marx and Engels' *The Communist Manifesto* and John Stuart Mill's *Principles of Political Economy*.	• Thomas Piketty's *Capital* and Shoshana Zuboff's *The Age of Surveillance Capitalism*.
Fiscal policy	• Independent Treasury Act of 1848 (off the gold standard).	• Maturing of crypto-currency. • Acceptance of China renminbi as global reserve currency (with the dollar, pound, euro, and yen).

Aspect of daily life	1848 (give or take a few years)	Roughly today
Value creation	• Decline of farming jobs vs. industry and services. • The California Gold Rush.	• The Global Code Rush. • "Code-generated value" outstrips that of physical products (top stock performers and valuation leaders are mostly digital).
Regulation	• Married Women's Property Act in New York (gave women property rights of their own and is seen by some as one of the first steps of the American feminist movement).	• GDPR, web content regulation in UK/Australia/EU, etc., Chinese *Social* Credit System, and more to come. • Rules constraining freedom of movement based on health status.
Migration	• Irish famine and diaspora, migration due to social turmoil (leading to 1848 becoming known as "the year that created immigrant America").	• More than 79.5 million displaced people in 2020, more to come. *The Age of Surveillance Capitalism.*

Western world, but the consequences of what was changing weren't readily apparent and wouldn't be for decades to come.

Perhaps it's a source of hope that we've been here before. Good, smart people came long before us, and while they may have wrestled with electricity or steam rather than sensors and software, the issues were pretty much the same. The machines may be new, but the path is not.

We shouldn't overlook that the turmoil of the late 1840s had some disastrous consequences. The American Civil War started in 1860, and World War I — the first truly mechanized war — started a few decades later in 1914. This begs the question: Are we approaching nirvana enabled by new technology or a global catastrophe? Our comparison lays out spooky similarities between the era of *Les Misérables* and our own miserable times.

The first time we did a trial run of the end-to-end narrative of what became our second book, which included a look back through history, the *very first question* from the audience was, "Every time in history we've seen this kind of disruptive shift, there's been a major war. Are we going to have a war?"

The response we gave then is the same one we give today. "What makes you think the war hasn't already begun?"

War has already been declared

Whether it's a spear, bow and arrow, catapult, machine gun, stealth bomber, mustard gas, attack helicopter, or nuclear weapons, from the time people started whacking each other with rocks and sticks, technology and fighting have been conjoined.

The ultimate exercise in hard power has *always* been monstrous, so what's new? Computing power has been used for fighting since the 1930s when it helped aim torpedoes. Later, the ENIAC — considered the first electronic general-purpose computer — was used to replace human "computers" (which is what they were called) to calculate missile trajectories in the World War II era.[5]

In our modern times, conflict is moving at *Blitzkrieg* velocity from metal to digital — drones, robots, and disruptive code. We're shifting from bullets to bytes, or as we pointed out earlier, from MAD (Mutually Assured Destruction) to MADD (Mutually Assured *Digital* Destruction).

In *Machines*, we said, "hybrid is the new black," meaning that businesses should extract value from *both* physical and digital sources. Now the same thing is happening to hard power. Digital conflict and physical conflict are like two sides of the same coin. One is not legal tender without the other.

Most of us (thankfully) lack firsthand experience with war, yet it's safe to say we have a media-informed collective vision of what we think war is like (thanks to media and movies like *Saving Private Ryan, Platoon, All Quiet On The Western Front, Full Metal Jacket*). However, the invisible tech war being waged with sensors, satellites, and software has yet to enter our collective understanding of "real" conflict.

Real shots will still be fired, so anyone picking a fight without preparing both bullets and bytes will end up thrashed as badly as the soldiers sitting in Maginot Line bunkers while the *Luftwaffe* simply flew overhead.

Consider some advice from Glenn S. Gerstell, former general counsel of the US National Security Agency and previously a member of the National Infrastructure Advisory Council:

> The digital revolution has urgent and profound implications for our federal national security agencies. It is almost impossible to overstate the challenges. If anything, we run the risk of thinking too conventionally about the future. . . . We must prepare for a world of incessant, relentless and omnipresent cyberconflict — in not only our national security and defense systems (where we are already used to that conflict) but also, more significantly, every aspect of our daily and commercial lives.[6]

Again, this may sound like science fiction, but it is, in fact, today's *non*-fiction.

- Malware called *Stuxnet* was allegedly developed to inhibit Iran's nuclear weapons program. The code was left on a USB drive that was dropped in a parking lot at an Iranian nuclear site. Someone plugged it into a production system computer by accident, and the virus spread. First, it tricked the monitoring systems into thinking nothing was wrong. Then it sped up nearly 1,000 centrifuges — perhaps being used to purify weapons-grade uranium — to the point where they simply flew apart. Some regard this as the first "shot" fired in true cyberwar, and a harbinger of things to come.[7]
- Triton has been called the "world's most murderous malware." It can actually "disable safety systems designed to prevent catastrophic industrial accidents." It's been used on companies in the Middle

East, but industrial companies all over the world are now being targeted.[8]

- We can glimpse Ground Zero of the endless digital war in Ukraine. In two months alone, there were 6,500 cyber-attacks across 36 targets — power plants, banks, transportation systems, and more. (That's a rate of 39,000 attacks in a year.)[9]

The list of lunges, parries, and *ripostes* gets longer each day as good and bad actors (it depends on where you're sitting) launch code at each other to disrupt power grids, industrial equipment, computer systems, mobile devices, data centers, oil refineries, and more. (These are just the ones we know about, the ones who got caught.)

So, now that we see we already have a war, what should we do about it? First, it's time to stop viewing conflict through the lens of the Greatest Generation. Tomorrow's war is already declared, already started, going on all around us. Advanced Persistent Threats are aptly named — they're technologically very advanced, looking for cracks in our digital castles persistently, and very threatening. But they're beneath our daily awareness because wars of the present just don't *look or feel* like wars of the past.

The roles of Robert Oppenheimer, Dwight Eisenhower, and Sir Bernard Montgomery are being played by unnamable leaders from Fort Meade, Maryland, (US Cyber Command and the NSA), Shanghai (People's Liberation Army unit 61398, specializing in cyber war), and Moscow (cyber warriors of GRU). The role of Audie Murphy is played by Edward Snowden. The role of Alan Turing is played by hacker kids coding in basements and garages in Jerusalem or Silver Spring or Shoreditch.

It's time to reset our ideas about what conflict will be tomorrow. Just because there may be less physical carnage doesn't mean there aren't fierce battles taking place, right now, inside our internet cables, media channels, and devices. We trivialize MADD at our peril.

Second, most countries are investing in how wars *were* fought, rather than how they *will* be fought. For example, a single modern aircraft carrier costs around $13 billion (not including research and development), while the entire public US cyber defense budget request for 2020 was around $9.6 billion.[10]

We're not suggesting that all investment should swing away from hard power to digital power. However, now that cyber war *matériel* can — in many cases — be more effective than metal war *matériel*, at a fraction of the cost, societies should question the investments being made in fighting last year's war (or even last century's war).

Finally, and perhaps most importantly, the goal, of course, is not to fight more efficient wars; it's to lower the requirement for fighting and minimize the cost to as close to zero as possible. Our goal must be peace (without indulging in the *naïveté* of thinking we can ever be fully conflict-free). Can we allocate less treasure to defense (and offense)? Can we reduce our collective needs for standing armies? Can we use new technology to inhibit bad actors without firing a shot, launching a missile, putting people in harm's way? Perhaps this shift from purely physical war to hybrid war gives us a unique opportunity. Perhaps now is the time to be open to the hope that we can use tech to take real steps toward peace through something other than bigger guns.

In the end, there will be vast differences in how each society responds, but a healthy future requires us to recognize that hybrid digital and physical conflict is an entirely new monster that must be first recognized and then faced head-on.[11]

Given the obvious parallels between 1848 and today, and the rise of a silent festering war that may not scare the horses but scares the bejesus out of us, there is an understandable temptation to run for the hills or simply (metaphorically) duck and cover. Our societies, riven by machine-created inequalities and human-created frailties, are suffering from stresses and strains that once again require us to marshal our better angels to ensure that evil doesn't flourish, and do so with the lightness and brightness that inspired John Lennon to say, "Everything will be OK in the end. If it's not OK, it's not the end."

6 A MANIFESTO FOR TAMING THE MONSTER

In which we propose a series of provocations about how we can renegotiate our "terms of endearment" with disruptive technologies — and ourselves — over the coming months and years. Our response must help us adapt to a decade that will be shaped by both technology and the pandemic, social (r)evolution, climate change, and more.

Can we simply turn off tech? Disconnect? Go dark? Maybe, but it's not easy. For many of us, it's simply not possible. (And what's the fun in that?!) What we can do, *must* do, is reflect on how we want to engage with tech, decide what we want for our societies (and ourselves), and then act accordingly. Maybe we can't turn it "off," but we sure as hell can turn it down!

Throughout the book, we've dropped crumbs and seeds of our initial ideas on how to begin to tame the beast. We'd like to leave you with our unequivocal list of declarations about what we *must* do to begin to tame the Monster, leaving us with healthy jobs, lives, and societies.

Over the years we've spent working on this book, it's become increasingly clear that our relationships with the technologies of our modern world have created a monster, because the machine — the thing we are building — is still incomplete. It's like we've built a car, but we're only halfway done. We haven't finished the brakes; the steering only works intermittently; and there is no navigation system installed. To tame the monster we've built so far, we need to first recognize it's not complete, and then we need to finish what we started. This manifesto is almost an assembly manual for building the rest of the machine.

Our goal, you'll recall, is to spark dialogue, not provide a complete answer to the ultimate question of life, the universe, and everything, including how to tame the Monster.[1] After all the miles traveled, research studies conducted, client meetings held, books, papers and articles written and read, and years of trying to understand what to say to help us craft and embrace a hopeful future, here is our manifesto for living with our creation.

This manifesto is not passive; it's a call to action for all of us. It's also incomplete. Each declaration could be the subject of a separate book — or 10 books! Our hope is that this curated list of provocations will help set a course that each of us — as citizens, workers, family members, and individuals — can follow.

I. Co-author new rules of the road

 Industrial-era regulations, practices, and laws have failed to ensure the health and well-being of our society and must be modernized to withstand new threats and enable new opportunities.

 The velocity and direction of the next phase of the digital economy will be driven by revised laws, policies, and regulations for our new machines: net neutrality, privacy, patent and IP law, taxation, data protection, industry regulation, AI ethics, labor laws, health data laws, job licensure, sharing-economy regulation, etc. Even if it sounds mind-numbingly dull, this must happen. It's up to us to reset the governance structures of our new machines.

We've always known there was a dark side to our increasingly digital lives. "Fake news," election tipping, and black markets for personal data are all branches from the same rotten tree. Painfully common bad behavior is driving **We The People** to collectively say, "Whoa! Have we gone a bit too far?" Just a few examples:

> • **Patent and IP rules.** The US Supreme Court's *Alice* ruling makes it difficult/impossible to patent software or business methods that execute an abstract idea or algorithm (e.g., a platform that manages escrow accounts).[2] Rulings like *Alice* do lower the number of nuisance intellectual property lawsuits, which is great, but as law firm Morningside IP noted, "It leaves many of us wondering: how long can a patent system that was developed to handle traditional inventions continue to function in the 21st century without a major overhaul?"[3] IP battles are already flaring up. Apple, Google, and others have already been fined billions for anti-trade practices, and the regulators are just warming up.[4]

- **Industry regulations.** Euro banks are moaning about the tech giants leaking into their space and positioning to use regulation as a wall.[5] And they may need a wall. Amazon and JPMorgan Chase are working together to figure out how to offer bank accounts.[6] Apple and Goldman Sachs are working to partner to offer a consumer credit card.[7] The sharing economy seems like something Marx wrote about in glowing terms, but even in progressive Denmark, Uber is illegal.[8]

- **Too big to fail?** The companies now "too big to fail" aren't banks at all; they are the Code Economy tech giants. Just five companies — Apple, Amazon, Google, Microsoft, and Facebook — now account for roughly 20% of the total worth of the S&P 500.[9] Apple is worth roughly $2 trillion, and Facebook, barely impacted financially by the Cambridge Analytica crisis, is worth $750 billion. If one of these fails, gets broken up, is hit with significant fines, we can expect a seismic financial impact.[10] (A round of Maalox please!)

- **That data is me!** The General Data Protection Regulation, ramped up in mid-2018, regulates how the personal data of all EU citizens must be treated.[11] And its implications extend beyond Europe. All multinational companies — not just the digital natives — must figure out how to more responsibly manage their code. <*OK, yawn . . .*> But, if they don't, it will be more than a slap on a few French-cuffed wrists. Penalties can be up to "4% of the worldwide annual revenue of the prior financial year."[12] For a large company, that could be billions in fines. <*Awake now!*>

- **Driverless cars stuck in park.** Even though the technology is already working amazingly well, "driverless" cars in the US may be speeding toward Constitutional gridlock.[13] The US Congress is moving toward allowing Level 5 autonomous vehicles, but this is heading into the quagmire of a federalism debate against state abilities to set driving regulations. Similar to Uber, driverless vehicles could work in "my" state or town, but not "yours."[14] The Sartre-esque sign at the border could well say, "No Human Driver? *Huis Clos*."

- **Taxing tech.** Tax codes are not immune to this shift either. The European Commission is working on a proposal to tax tech giants based on where they generate revenue, not where their headquarters are located.[15] If you are sitting in Spain and you buy an iPad on Amazon.com, Amazon may have to pay tax in Spain, not just in Luxembourg (the company's regional headquarters with friendly tax rates). This will add up — fast.

Just as the Magna Carta, gerrymandering, the Code of Hammurabi, and banking regulations have shaped lives for decades or centuries, the policies and protocols we're cobbling together today for the digital economy — our Declaration of *Inter*dependence, perhaps — are crucial to shaping how we engage with each other, work with each other, and even argue with each other for generations to come.

Regulation has been — and still is to many — a dirty word, but we've argued here that we deserve better technology governance. We simply *must* find a way to create and manage new "rules of the road." The real challenge is: How?

Like it or not, lawyers, politicians, courts, and policy wonks are quietly — with our implicit or explicit support — rewiring the rules that will shape our lives for years to come.

We fully recognize that the bad behavior we outline in this short book has actually been confined to a small number of companies operating in a few specialized parts of the technology arena. So, legislators and voters need to be able to distinguish between that part of society/the economy, and others.

Technology companies haven't abused trust in supply chain management, in medical device manufacturing, in deploying smart connected cars, in creating entertainment options that a decade ago would have seemed mind-boggling. Technology is at the heart of flying, of shopping, of communicating with our friends and family and colleagues, of the healthcare we receive. And how do people feel about those technologies? Fine — better than fine in many cases.

Senators, congressmen and women, citizens, members of parliament, company leaders, members of the commission — don't forget what is at stake here. Let's stop bad behavior and bad actors where we can. Let's put the rules of the road in place for the information superhighway. But let's not overdo it and regulate things that don't need to be regulated. The next waves of growth and opportunity — for everyone — will come from smart technologies that continue to unfold, smart management by those with a hand on the tiller, and smart use by informed consumers. This mix of outside-in and inside-out regulation must be a central feature of the next phase of our disruptive times.

Nothing less will do. Technology has a vital role to play in our response to the COVID-19 pandemic and in building a more civil society. It is even more important now to get this balance right than it was a mere few months ago. The stakes were enormous back in January 2020. They're astronomical now.

As we said in this book's opening section, nobody has all the answers, and reimagining tech governance comes with a host of challenges. What we *do* know for certain, though, is that we *must* address taming the Monster with ideas, regulations, laws, and daily practices that are as powerful as the challenges we face.

II. Govern technology by community

 Technology is too important to be left in the hands of technologists. *All* must participate in active communal regulation to serve the greater good. Participation is part of responsible citizenship.

 Intimidated by the complexity of technology, "civilians" have tended to let the "nerds" get on with things on their own. How did that turn out? Not so well. Technology must be regulated by technologists and a diverse cohort of non-engineers who probably can't spell "Java." Technology experts may not like the sound of this, but we will get used to it for a healthier future.

A central feature of our disruptive times has been the absence of regulation. Indeed, the irony is that the Monster was born and grew in an era in which the real demon was the heavy hand of government oversight. Between Ronald Reagan — *"Government is not the solution to the problem. Government is the problem"* — and Gordon Gekko — *"Greed is good"* — technology has developed in a vacuum of serious engagement from those outside of the technology church.[16]

But that era is over. Merely having regulation isn't enough. As Shoshanna Zuboff notes in *The Age of Surveillance Capitalism*, it's just as critical to answer: Who gets to decide?

Our view is that technology is now too important to be left solely in the hands of technologists. Auto makers originally fought against seat belts. Tobacco companies fought against regulation. The list goes on, and the same thing is happening with tech. Self-regulation is being exposed, again, as a myth, a wish. Now we're being reminded of Douglas Bader's brilliant insight (instrumental to winning The Battle of Britain) that "Rules are for the obedience of fools and the guidance of wise men."

So far, the digital revolution has been largely advanced by engineers and computer scientists. Someone had an idea, built something cool, and then put it into the market. In the Facebook/Uber/Google mythos, this is then followed by unimaginable wealth, yachts, and a *Vanity Fair* photo shoot.

We wrote in *Code Halos* about the commercial power of personal and business data. We suggested that business leaders should be proactive about ensuring trust and transparency in how they use data to make better products. While many did, the siren song of making *beaucoup* bucks drove many others to monetize data without delivering value back to the people — us — who created and shared that data.

It's easy to wag our fingers at Silicon Valley and Silicon Fens and Silicon Shenzhen, but actually we share the blame. Yes, perhaps "they" should have been more aware and open about what "they" were doing. But likewise, it would have also been better for all of us if "we" had maintained an awareness of what we're doing as we invited others to scrape our data to target ads and content in exchange for an endless scroll of Pinterest boards and TikTok clips.

While professional people in every field have expertise that ought to be respected and considered, leaving critical decisions solely in the hands of "the experts" is a surefire way to have your opinions and convictions marginalized. If you're like any other healthy adult, you really don't want someone else making decisions for you, so you have to show up and be heard. If we don't, we can't complain about the outcome.

Of course, we don't want to throw the baby out with the bathwater; a course over-correction that dilutes innovation, stifles competitive behavior, and handicaps disruption across all technologies would be an historic mistake that could undermine the West's medium- to long-term future. At a time when the future of work is being built around the world and rests on the new tools and techniques we see emerging in the businesses in which many of us work, a turn against technology runs the risk of waving this future goodbye.

To balance out differing — and sometimes competing — motivations, it will take a village to truly shepherd the Monster. Civilians, amateurs, disinterested parties, the consumer, the dispossessed — all of these voices need to be represented at the regulatory table, sitting next to the developers, the AI wizards, the "arms dealers," and the capital owners. Tech/AI/ Monster regulation should take inspiration from how atomic energy is managed — the board of the UK's Atomic Energy Authority is split 50/50 between scientists and executives from within the industry and heavyweights from outside.

But accompanying this outside-in regulation, there needs to be recognition that the most material impact that will chart the future course of technology won't emerge from just politicians but also from us. Ultimately, it is up to us to save ourselves.

III. Apply The Golden Rule in cyberspace

 In order to avoid fueling anger, contempt, and cruelty, our online interactions must include the same level of compassion we are taught to offer in person.

 Treating others with compassion is simply not as easy as we might hope. While the philosophical roots of the Golden Rule go back 3,000 years and appear in nearly every religious tradition, we still haven't got it all right. The perceived anonymity provided to us by the screen — the ubiquitous screen — invites us to unleash our baser instincts. Taming the Monster requires us to recognize that our online interactions must include the same level of compassion we know we should show in person.

We know we should limit our screen time. We know we should balance our chase of low online prices against the damage this does to employment in local stores. We know we must judge whether the creation is worth the destruction in Schumpeter's equation.[17]

And still we don't, at least not yet. Not enough of us anyway.

This inside-out regulation we know we need speaks to a moral renaissance in the culture — a culture made up of the trillions of micro actions that we, the collective *we*, take every second of every day around the world. If we like fake, sensationalist nonsense, if we write hateful comments in a tweet stream, if we just want the cheapest thing (with no thought for the externalities), then we will get doused in the septic tank of existential vulgarity we're getting now, and no amount of outside-in regulation will really bend the moral arc of the universe. It may sound trite to the cynic, but we actually *do* have to be the change we want to see.

It is this inside-out regulation — our individual, personal behavior in every screen-based interaction — that ultimately holds the key to taming the Monster before it tames us.

The Golden Rule has been known to us since time immemorial: *Do unto others as you would have them do unto you.* Some form of this edict is in virtually every theology because: (a) it's good guidance, and (b) it's hard to do. Our declaration here is that this is every bit as applicable through the Gorilla Glass screens that frame our lives as in the market square and your neighborhood. It is only through recognizing this basic truth that we can actually save ourselves and each other.

It's not easy, even with thousands of years of practice trying to get it right, but it really is that straightforward.

IV. Accept your role as part of the solution

 Recognize your individual role in creating the problems with technology. This awareness will illuminate the fact that we have the agency we need to save technology and ourselves.

 It's a popular pastime to blame and shame the leaders of social media and tech companies for the sins of the Monster. However, in the words of Pogo (from the Walt Kelly cartoon strip), "We have met the enemy, and he is us." The problem isn't "them;" it's you! It's Paul! It's Ben! It really is us! Taming the Monster requires our own awareness that we can't simply blame Big Tech innovators for giving us what we wanted, what we asked for, what we feed, what we won't turn off.

What's really happened in the modern world — when you strip away all the rhetoric and rococo flourishes — is that in the seeming blink of an eye (in reality the last 15 years), tech has gone from geek to chic to . . . dirtbag. The "PC guy" and the "Dude, you're getting a Dell" guy were replaced by Steve Jobs and the Web 2.0 guys (almost all guys) who took COBOL and router seeds and turned them into dating apps and streaming services. For a while this seemed cool, and everyone was down with it. The disrupters were given a pass, and nobody dared question the collateral damage of "disruption."

But then we started to notice that the cool kids, clad in Allbirds and Patagonia vests, weren't the fun-loving chillsters we imagined but were instead just the latest rev of get-rich-quick, snake-oil salespeople that we thought had been exposed (and killed off) by Arthur Miller and David Mamet.

It embarrasses the authors to say we bought into this perhaps a little too much and cheered a little too loudly for the new masters of the universe. We saw both potential and peril, but we were betting against peril. *Code Halos* evangelized the idea of data-based mass hyper-personalization. We wrote white papers about the upsides of becoming a "social organization." We wrote *What To Do When Machines Do Everything*, which was overall bullish on technology but also predicted that AI would destroy 10% of current jobs. We took space in a coworking facility and beat the drum for a We generation.

All of that we did in good faith and a belief/hope that the path of progress was winding, went through some rough neighborhoods on the way, but in the end landed in a good place.

But then we started to realize that data-based mass hyper-personalization was creating the aforementioned Monster. That social technology was destroying a generation of kids — and democracy. That AI was pumping steroids into the winner-takes-all society that was seeing ordinary people jettisoned from the promise of a middle-class life. That coworking was really just a way to monetize folks with a bad case of FOMO.

We started to see that the innovators we had lauded as heroes were actually far from heroic. It would be rude (and potentially litigious) to name names. But we imagine you can imagine who we're talking about. . . .

No, these tech savants weren't always heroic. In fact, some of them are seeming like JADs. Just Another Dirtbag. Not cool at all. They'd been given license to upset apple carts with financing sourced from some of the dodgiest people in the world, people who were being promised fat returns for nothing. People benefiting from r>g. The disruptors didn't have to build real businesses; all they had to do was create scale (get from 0-1), get through the IPO window, make their financiers even richer, and then, well, the rest would take care of itself. *Once you've got FU money, who cares what happens?*

In short, tech became the main ingredient of a toxic brew, and so was it any wonder that ideas like *Sunflower* — in our earlier screenplay treatment — went from being cool and fun to odious in the blink of an eye? Is it any wonder that the beauty of a young idealist's vision mutates into the ugliness of a machine whose only purpose is to make as much money as quickly as possible?

Is it any wonder that weird, long-term, high-risk/low-return ideas (which have the potential to positively change the world) have almost no chance of getting funding? Is it any wonder that the financial intermediaries — that in aggregate act akin to a single-ecosystem algorithm — have learned the recipe of short-term success and nudge their algorithms to favor ideas that are already familiar? Is it any wonder that every new app/URL looks and feels the same? Is it any wonder that harvesting user data to sell ads is now the business model of the future? (As an aside, does anyone remember that people left TV for the internet to get away from advertising in the first place?)

As futurists — people whose passion and job is to find the unevenly distributed future and share its glories — it pains us to say that the future is being ruined by money and by people who are chasing it with no holds barred. This is true, even if Big Tech plays a role in dealing with the COVID-19 pandemic.

We've got to wake up. We've got to stop the madness of the Faustian bargain of funding for growth at any cost. We've got to place long-term bets that don't rely on advertising-based business models. We've got to favor history-making moonshots over instantly forgettable lowest-common-denominator slam dunks.

We have got to spend our dollars with companies that don't monetize our data in cavalier ways. We've got to realize that there's nothing as expensive as a free lunch, and paying directly for things typically makes sense in the long run.

Blaming others for the Monster is weakness masquerading as reason. It may be fashionable to screech at the tech oligarchs, the digital masters of the universe, our modern-economy robber barons, but it's time to look in the mirror. We've got to become aware of the fundamental truth that the choices we are making are feeding the Monster.

Sitting on the sidelines is for wimps. Don't. Waiting for "someone else" to figure this out for you — *your* family, *your* company, *your* country — is a mistake. If you're a member of a democratic society, you have the right (obligation, really) to exercise authority over how you manage tech, use tech (don't be a troll), and — critically — how you participate in the democratic process to govern tech.

Wake up! Wake up! Face the truth! Wake up!

V. Don't give up on loving tech

 Any powerful tool can be used for ill or good. The new technology that makes up the Monster is no different. We must make peace with it, control it, and nurture it to better our society.

 Fire, electricity, and even water can both destroy and create. The same is true for our new machines. We don't hate fire for burning down a building, or water for flooding a field. We invent tools to control fire, control water, make them productive, not just destructive. The first step is to resist giving in to greed. The second is to resist giving in to hate or fear for the damage technology can cause (and has caused).

It is getting harder and harder to see an acceptable face of our current version of capitalism. Without countervailing forces in the world — a code of ethics, regulation, modernized social conventions, the critique of Marxists backed by tanks and nuclear weapons — Capitalism 1.0 reigns supreme and year on year becomes more unfettered in its ability to create wealth, at greater scale, more quickly, at any cost, and in Piketty's formula, in ways that deliver more and more of that wealth into the hands of the existing owners of wealth. If capitalism is so great, its critics are asking, why does it have to be bailed out by socialism every decade, as it is again as we write, in the wake of COVID-19?

Tech is core to that idea — the fulcrum of growth in a world grown stagnant under the weight of its wealth. Tech and late-stage capitalism are melded together in a way that Stewart Brand and Woz and other members of the Homebrew Computer Club would have never imagined.[18] Apple was meant to "stick it to the man." Now Apple is the Trillion-Dollar Human.

What's wrong with tech is that it has become a handmaiden to money rather than a conjurer of new and better worlds. Tech has always been the wellspring of the "better" — better health, better food, better security, better education, better entertainment, better transport. But now, tech can only disrupt if the disruption makes money — to hell with making anything better. Tech — the tech we love, the tech you love, the tech in the best sci-fi — is being soiled (nay, spoiled) by the greed of the past. Tech has kowtowed to capital in a way that can't and won't end well. After all, being a handmaiden never has a happy ending.

Beyond polluting our wellspring of capital, we've shown how the Monster is hurting our minds and corrupting elements of our society many of us hold dear. Technology is acting like your relative or friend who keeps having too much to drink and wrecking social situations — endearing at first but then harder to love and defend as time goes by.

People!!! We've got to change this. Before it's too late. And it starts today with YOUR next click.

To actually make this work, we have to get back to the core belief that improving our tech is feasible. Perhaps it sounds just too far out, but getting back to this belief is only possible if we have faith that tech is still worth believing in, still worth caring for, still worth tending to, that it can be more than a Monster to be tamed or — worse — snuffed out.

Just as we put rocks around a fire pit, or create traffic laws, or encode our values in a constitution or book of laws, the steps we take to govern new technology to benefit society, rather than harm it, are all within our human hands. On the job, fretting over our home privacy settings, or in the voting booth, each of us must enter this next phase of the Fourth Industrial Revolution with a sense of responsibility about what we put in place to govern our relationships with technology and, therefore, with each other.[19]

This is the great task ahead, a task at which we must not fail — a task great enough to inspire this book and all the anger and energy and hope and frustration we trust you have sensed spilling through these pages. If our immoderation has insulted you, we apologize. If it has inspired you, our work is done.

VI. Treat your data like your reputation

 Intemperance and carelessness with our personal data damages our lives and society and must be avoided. We must each act to preserve the value and sanctity of our individual code while still participating in our modern world.

 Our data — tweets, account numbers, images, location, voices, medical information, emails — creates a penumbra of our characters, our selves. At the same time, centralized data control is the taproot of winner-take-all internet businesses. Some are trying to wrest complete control back from the digital oligarchs and hackers, but that is unlikely. Expecting politicians or business leaders — each with their own motives — to care for our data is folly. It's up to each of us to recognize our personal data as a precious asset, like our reputation, and treat it as such.

As we have shown, most of our personal data is gathered via our uninformed consent, but a vast amount of useful information is harvested by bad actors, people, and organizations who recognize they are taking data from us that they fence to others on dark web exchanges.

According to SpyCloud researchers, in 2019 alone, nine billion credentials — our username and password information — were "spilled" from 270 million users.[20] ("Spilled" makes it sound like you merely knocked over your cocktail at the bar, but it really means you and several million people you don't know just had your bank accounts and Social Security numbers opened to some actual bad guys.)

Shape Security found that credential "stuffing" — using that stolen data — accounts for 80% to 90% of retailer e-commerce logins.[21] In 2021, the estimated impact of cybercrime alone, not including ruptured societies and nudged elections, will be around $6 trillion (more than the 2019 GDP of Japan).[22]

There are literally countless tactical steps we all know we should take on a daily basis. Install security patches; consider and check before posting personally identifiable information; avoid blithely clicking <Accept> on every legal agreement; double-check that email address or URL to prevent biting into a phishing hook. But all these small steps start with a shift in mindset that turns into a regular practice.

In the days of push-button door locks, locking our cars was a manual and mechanical procedure. Ensuring that the Ford Country Squire or Vauxhall Victor stayed where we left it was an unconscious action. We didn't count on or require anything other than the car company to make the door *lockable*. In a sense, we worked with the car manufacturers to protect our assets.

We can apply this same philosophy to our personal security by taking three specific steps:

1. **Demand the tools for data control.** We must hold the technology toolmakers responsible. Every company that requires our precious data to do *anything* must fully and completely own the responsibility of allowing us to take easy steps to curate and protect our own data.

2. **Use those tools to take control.** Simultaneously, we must recognize that no security system works without us doing our fair share, exercising our personal agency over our own security and privacy. There is no way any of us can single-handedly combat the legion of bad actors intending to harvest value from our data. What we *can* do, and *must* do, is take those tactical steps and make them part of our daily habits — not unlike brushing our teeth, locking our car doors, or arming our home security systems at the end of the day.

3. **Recognize data protection as a national security challenge.** Government officials and policymakers must feel responsible, and be held accountable, for spearheading the defense and offense against attacks current and pending that have the potential to disrupt life and liberty. It's straightforward to point at a 100,000-ton aircraft

carrier and say, "We're using this to keep you safe." Perhaps it's less satisfying to point to a drab room full of white-hat coders and say the same thing, but we must. Ransomware attacks, DNS hacks, complex phishing expeditions, and more all add up to a digital *Blitzkrieg* that requires a coordinated international, national, and state-level government response.

Whether you're making policy, leading a business, or simply managing your home, it's tiring to feel under endless attack, but it's time to recognize the endless war against hackers, both foreign and domestic. That means taking persistent steps to stop inviting the Monster to become even more powerful and disruptive.

VII. Fight against the relentless brain hacks

 Deliberate use of technology to impact our behavior without consent and awareness must be viewed as a clear and present danger to society.

 Tech gives us endless stimulation, a little drip of dopamine with every click, tweet, swipe, and like. It's time to admit the psychological impact of the Monster. Any use of technology + psychology to addict or exploit us without our actual consent or knowledge must be recognized as a violation of trust and ethics. This practice must be rooted out and shut down with regulation, law, and social convention so that our activities, thoughts, and emotions are not hacked, tracked, and manipulated every time we log on.

"You are hooked. Our kids are hooked. I am hooked."

Say these words out loud, and *mean* them, and you've just taken the first step toward kicking your own addiction to new technology.

Recognizing the problem is the first step to making the daily decisions *not* to reach for the smartphone because you're afraid of missing something, to focus on your kid when she's trying to talk to you, to avoid hitting refresh on that latest Instagram feed that is *not so different* from the Instagram feed of five minutes ago. . . .

We all get this, and we've discussed it earlier here, but the big question now is: What do we do about it?

Start thinking about our craving for new technology as being like our craving for sugar. Almost all of us love it, but sugar would likely be illegal if it came on the market today and was subject to government oversight because it's harmful (diabetes, obesity, etc.) and addictive.

The comparison between technology and sugar is not such a leap. Even Facebook executives are comparing the addictive nature of their platform to the addictive nature of sugar. In an internal post from 2019, a Facebook executive asserted, "While Facebook may not be nicotine, I think it is probably like sugar."[23]

If addictive apps are as bad as sugar, but not as terrible as nicotine, how bad is that? Studies show that sugar releases opioids and dopamine, indicating addictive potential. (Sound familiar?) Some researchers even advocate for sugar to be considered a controlled substance. British scientists found a remarkable overlap in response to sugar vs. illegal drugs like opioids.[24] The Centers for Disease Control and Prevention (CDC) suggests that in the US alone, sugar contributes to the obesity-related deaths of as many as 365,000 Americans per year.[25]

Of course, social media is not causing the same level of morbidity and mortality as cigarettes or sugar. But as we've seen, the Monster is absolutely not benign. It is addictive, and that addiction impacts our minds, the flow of trillions of dollars of capital, and our society.

Recognizing the issue is the first step, but to *really* tame the Monster, we each need to act. There are two actions we must take.

Channel your inner Marcus Aurelius

Most of us can have a couple glasses of wine with friends, but we know we shouldn't knock back a liter of bourbon daily. It's fine to pop some ibuprofen for your aching knees, but if you're using a Pez dispenser for synthetic opioids to get through your day, it's a problem that warrants immediate help.

The same is true for technology dependency. When we find ourselves too often falling down the rabbit hole of "doom scrolling" through the news or seeking up-to-the-second validation from social media, we need to call it what it is: a problem. Then we need to handle it. Read a few more paperbacks and a bit less TikTok. Don't feed the Monster so much with your every thought or image. Check Facebook/Instagram/your messaging app several times a day if you want, but not every two minutes.

Of course, this is much (much!) easier said than done, and it's personal. Both our families are fighting the same battle against being too jacked in. But there is no cavalry coming to help, so we can't ignore the fundamental requirement for personal agency. A tech detox starts with us, and there are no shortcuts.

We are *not* saying the only way to keep the Monster out of our heads is to go cold turkey, turn it all off, melt down your laptop. Unless you have an actual addiction, most of us don't have to turn it **all** *off* (more on that later), but it would be better for most of us to *turn it down*.[26] We are saying that each of us must cowgirl up and exercise control over the poison that compels us to engage.

This may sound a bit simplistic, but recognizing tech as an addictive substance, and admitting we are all susceptible is the first step toward understanding that it's in our power to hit the <pause> button.

Demand an easy-access "off-switch"

If you're starting to blame yourself, maybe feel a little bit guilty, hold on a minute. In many ways it's not a fair fight because, as we've shown, some of the smartest people in the world are linking technology and psychology to keep you hooked. Unimaginable capital is funding software engineers with an extreme level of mental amperage — right this minute — aimed at creating technology that is *intended by design* to be like a huge accessible bowl of sugar cookies at a toddler's birthday party or other temptations (legal or not) at a college-age rave. There's a reason cocktails are not served at AA meetings.

We've discussed overall regulation throughout this book, but there is one other tactical step we can all take. We must require an *off-switch*.

In *Code Halos*, we suggested that to build and retain trust, technologists and business decision-makers should give your Code Halo a delete button. Our view was — and remains — that individuals generating data for someone else's business benefit should be able to easily opt out of that commercial arrangement. This may sound completely reasonable, but as we've shown, the economic imperative regularly overwhelms doing the right thing.

If you want proof, try this "simple" Monster control experiment. (We will use TikTok for this example, but take any other app you care to try.) We'll make four assumptions.

1. Your home has Wi-Fi that you fund via a service provider and a home router that you own.

2. Your home has devices — laptops, mobile devices, etc. — that you also own to be used by you and others. (Any parent will recognize this setup.)

3. The TikTok app is installed on one or more of your devices.

4. You wish to make TikTok inaccessible in *your* home on the systems *you* own.

Now try to turn TikTok off. We don't mean simply uninstalling the app from your devices. Try to make TikTok unavailable to anyone on the systems *you own and are paying for*.

We'll wait. . . .

Spoiler alert! It's not impossible, but it's close. You'll spend hours looking up instructions. You'll set up an OpenDNS service account and configure your router to go through the OpenDNS IPv4 addresses. You'll try to block the entire series of at least 13 separate TikTok domain names required to shut it out. And more. It's an *oh-my-god* level of technical steps.

Maybe you can excise TikTok from your home, maybe not. The point is that along every step of the way, others have collectively decided to make it as difficult as possible for you to tame the Monster in your own home, on your own equipment, on internet access *you* are paying for.

We work with some of the top business leaders in the world, and we have yet to meet anyone who explicitly intends to be creepy or evil with personal data and information, but that is the unintended end result. Unless we require software developers, company leaders, internet service providers, and governments to give us an *off-switch* (one that doesn't require a PhD in computer science to use), this is what we should expect.

All of this raises legitimate questions about individual freedom. None of us want regulators inhibiting or monitoring us as we scroll Instagram. We don't want to control the tracking by getting more people to track us! It would be madness to invite that kind of mafia tactic for "protection."

Our view is that freedom to use technology according to our whims sounds great just until the technology gets creepy and evil. And it has. Actually being "free to use technology however we want" requires that we are *not* hijacked, that we are *not* manipulated by the most powerful technologies created by humans. These systems are designed to be addictive, and *that must stop.* Hacking our brains without our full awareness m*ust* be treated as an intolerable violation of trust and ethics.

Fictional monsters throughout time have been brought down by the collective action of the populace. We need that now. Slowing the Monster's assault on our minds is easier said than done because so much money is involved. Our next steps involve both personal agency and simultaneously holding technology builders and providers responsible for giving us the tools to put a stop to the profoundly unhealthy impact the Monster is having on our minds. Anything less will ensure we continue to have monsters under our beds, in our pockets, in the walls, in our cars, and in our heads.

VIII. Modernize the authority over technology and capital

 Regulations governing capital have not kept pace with the power of the new machines. To ensure innovation, safety, freedom, and peace, our rules and practices related to the flow of money must be modernized.

 Our new machines have upended centuries-old assumptions about means of production, investment, IP protection, tax, trade regulations, antitrust, capital markets, private equity, stock exchanges, and so much more. We must all actively participate in making informed decisions about the relationship between capital and technology to protect wealth, innovation, social security, and even our personal happiness.

We have shown how capital — cash, filthy lucre — is really calling the shots on how technology is shaping our jobs, lives, and future. Controlling "the money" warrants a bit more focus.

Companies ranging from funded startups to those members of the Trillion Dollar Club all bear some responsibility to increase shareholder value. We've shown that the most valuable companies in the world have value propositions based on software and *our data*.

The upshot is that basically every company in the business tech game is, one way or another, *compelled* to take as much of our data as possible — captured by any means necessary — and monetize it. Until we change the rules of capital, they have no choice. Where did these rules come from? Why are they so broken? What should we do?

The Code of Hammurabi is considered one of the first legal documents. It was written (carved actually) around 1750 BCE to codify some of the earliest regulations related to trade (among other things). Other legal codes emerged over time. Practices related to lending money, storing wealth, and even setting interest rates have been around, and regulated, for thousands of years. Double-entry accounting came into more regular use in Europe in the 1300s, and the first formal stock exchange was in the Netherlands in the relatively recent 1600s.[27]

In the late 1800s, long before the transistor was invented, US antitrust laws were put in place to protect our physical economy against restraint of trade, price fixing, cartels, and monopolies. We obviously still need these rules, but the regulatory logic flies apart in the digital economy, which operates with entirely different laws of physics.

Consider how much you pay to use TikTok or Facebook or Google apps. It's $0. Fine, but this completely overlooks the fact that the actual "price" we're paying is our attention to advertising and the data we're granting that is then monetized to help the platforms target and sell advertising. Other search engines and social platforms exist, but without a similarly massive and growing stockpile of user data, they simply are not as valuable, and they never will be.

Some disagree, and they have good points, but our view is that the laws in place today were designed for fundamentally different commercial models. Are these companies monopolies in the eyes of the law? What is the real "price" of their goods and services? Are platforms actually media companies? It's a good time to be a tech lawyer because the courts will be busy for years wrangling with (and complicating) these issues.

It's obvious to anyone watching testimony from tech leaders to the US Congress or the European Parliament that the pace of innovation is outstripping the velocity of the regulators, and this is allowing the Monster to wreak havoc and grow more like an invasive weed than a healthy business.

The monks, merchants, royalty, and bankers who pioneered our financial systems, and later advanced them through the Industrial Revolution and beyond, could *never* have conceived of a trillion-dollar company that didn't make anything other than software. Authors of the Sherman Antitrust Act of 1890 simply could not have imagined commercial models where the dollar price for a software-based service was zero.[28] Early investors and regulators could not have imagined huge valuations for companies that have effectively almost no net profit margin (or less), like Salesforce, Snowflake, Tesla, Uber, and so on. The clear conclusion is that our laws, practices, and even our business philosophies related to managing capital remain more aligned with industrial-era powerhouses like Bethlehem Steel (RIP), Sears (Chapter 11), and GE (dropped from the NYSE) than with Facebook or Amazon.

The entanglement of money and power isn't new, and as individuals, most of us can't do a lot about banking regulations. What we can do, and what we must do, in fact, is demand that our elected officials put in place modern policies and laws related to governing the relationships between money and technology and power.

Specifically, a healthy future requires us to modernize the rules to shut down covert surveillance-based capitalism and to create healthy mechanisms for managing capital in the digital economy, such as antitrust laws that reflect data-based commerce; laws reflecting the fact that social platforms are actually a new kind of "press" rather than a *tabula rasa* town square.

We get that this is a heavy lift. The prospect of listening to political candidates bloviating about antitrust laws from 1890 and stock market governance sounds as dull as dishwater. But we must not abdicate our role in sculpting the future of our society. This *matters*.

Maybe you want to sit this out. OK, but if you believe companies are capable of self-regulating, chances are you also believe in a flat earth, faked moon landings, and Elvis (P) living among us. The *only* way to tame capital is to change the operating rules of information and money with regulations, tax policies, controls on capital markets, and more. Throughout history, this is how we decreased smoking, broke up innovation-crushing monopolies, got seat belts and airbags, and grew wealth at an unprecedented rate. It may be tough, but we know we can do it, and we know we must.

One question that must be addressed head-on, though, is this: The Monster is (partly anyway) comprised of technologies and devices all with an <off> switch. We simply wouldn't have a Monster, nor would we need a manifesto, if we turned it off. So why don't we simply turn it *all* off?

7 OFF?

In which we debate the difficult but unavoidable question: Isn't the real solution to all the problems we've outlined in this book to simply turn the Monster off?

From: Ben   3:32 PM
To: Paul
Subject: I'm mad as hell and I want to turn it off . . .

Politicians are now climbing onboard the tech-lash bandwagon. Breaking up the tech giants became a serious idea in the 2020 US presidential campaign.

Facebook recently hired UK's ex-Deputy Prime Minister Nick Clegg to make the case that breaking up Facebook won't fix anything. This feels big.

From: Paul   3:36 PM
To: Ben
Subject: Something big *is* brewing . . .

Zuboff's book on surveillance capitalism pulls no punches and reads like a 704-page indictment against social media platforms, data brokers, and basically everyone. The digital saber-rattling between China and the US is getting terrifying.

When the right and the left agree on anything today, you know something important is happening, and they all hate the social platforms.

The Big Tech Dogs — Pichai, Zuckerberg, Dorsey, Cook — are all trying to avoid being called to the hot seat in DC or Brussels to explain what's happening.

From: Ben   3:41 PM
To: Paul
Subject: Hating on social media

Social media is now in *everybody's* dog house. It started out as a miracle, but today nobody has a good word to say about it. Maybe the coronavirus has taken the spotlight off the downsides of tech, and allowed Big Tech a chance to refurbish its reputation, but the underlying issues are still going to be here once we're past this moment of panic.

From: Paul   3:42 PM
To: Ben
Subject: Re: Hating on social media

Zuck sat in his dorm room and invented a fun way for smart kids at Harvard to meet cute and share cat pix.

He could *never have imagined* his app would turn into a tool for messing with our minds, elections, and money in just a few short years.

Quarantine drove us to jack in even more, and it was great, but I think we'll find that our collective tech-binge will leave us with a hangover.

Have we created a monster? Maybe we didn't mean to, but when you look at the whole picture — including social but also cloud, analytics, Zuboff's notion of capital — and the human toll, it's pretty clear that we have.

It's here, slithering around. The question now is really, what do we do about it?

From: Ben   3:55 PM
To: Paul
Subject: Turn it off?

Yes, but what's the real unasked question?

It's almost unnatural for tech-loving futurists who spend many waking (and non-waking) hours trying to look over the horizon to see what will happen with tech, money, business, and power.

We've all wished for it sometimes. It's so big and so small that it's hard to think about. It's almost unaskable.

Why don't we just turn it off?

From: Paul   4:56 PM
To: Ben
Subject: Re: Turn it off?

Turn what off?

From: Ben   5:01 PM
To: Paul
Subject: Turn it off?

ALL of it . . . Facebook, Google, Netflix, texting, Twitter, maps, LinkedIn, Amazon, Instagram, Snapchat, TikTok, Instagram, everything that feeds the Monster.

It's getting increasingly difficult, and even a bit uncomfortable, to balance the hopeful story about technology we've been telling for years with what we're seeing play out in real time. Of course, with the pandemic, and everyone being stuck at home, people are even more reliant on their tech, and getting even more addicted. But surely that just raises the need to ask, why not just turn the bloody thing off?!

From: Paul   5:03 PM
To: Ben
Subject: This is insane . . .

Come on man. . . .

From: Ben 5:15 PM
To: Paul
Subject: Is this insane?

If we can see that social media is messing with kids' heads, with their parents' heads too, with people's ability to be civil and decent, with relationships, with productivity, with "honorable" capitalism, with democracy itself, why don't we just walk away?

Why don't we recognize the error of our ways? That social media was an experiment, a period in time, a fad, a passing fancy, a mania, a bubble, a craze, a phase that we all went through.

Why don't we turn it off and go and do something better instead? Play football, or go for a ride, or make our sales quota, or read a book, or weed the garden, or run for office, or make some pasta — all without tweeting or Instagramming or Facebooking about it. . . .

From: Paul 7:18 PM
To: Ben
Subject: It's not ALL bad!

So we're saying the best response to the humans creating trillion-dollar companies, connecting people around the world, improving health, building communities, sharing knowledge, making transportation safer, helping people bank, and more — should be *switched off*?

From: Ben
To: Paul
Subject: Why not?

7:22 PM

We're also saying it's a Monster! Why don't we turn it off? We can! *Why?*

That's it. That's the question.

I don't have an answer.

From: Paul
To: Ben
Subject: Can't fight human nature

8:25 PM

Well, I don't think we can, and I don't think we should.

From: Ben
To: Paul
Subject: Re: Can't fight human nature

8:27 PM

OK....

From: Paul   9:35 PM
To: Ben
Subject: Re: Can't fight human nature

I just don't think we're wired for that.

At no point in history have humans ever walked away from a new technology and just said collectively, "Nope. Even though we know it exists, we're not going to use it." That's even more true now.

From: Ben   9:36 PM
To: Paul
Subject: Re: Can't fight human nature

I'm not sure that's true.

Monster

From: Paul   9:54 PM
To: Ben
Subject: Yet another history lesson

Name one thing!

I read a book called *Giving Up the Gun: Japan's Reversion to the Sword, 1543-1879*. In the mid-1500s, the Portuguese sent ships to Japan. It was largely a closed feudal society running on the *Bushido* code of power and honor.

They also didn't have firearms, which the Portuguese did. Those big *arquebus*-looking things.

The Japanese thought about it a bit, and they made firearms and even fought with them for a time, but then they decided that even though the technology was there for the taking, they didn't like it because it went against how they felt war should be fought. Their traditional social concept of combat was that it should be mostly close quarters — with swords — not standing at a great distance lobbing shots at each other. That was seen as less honorable. So they remarkably mostly hit the <pause> button and didn't seriously use firearms in combat until the mid-1800s.

But then they did!

Same thing in China. They had a small railroad built in the late 1800s but then decided they didn't like it; they tore it up and then sat on the sidelines for most of the Industrial Revolution.

And then they jumped back in full force and became the world's largest industrial economy.

Same thing with nuclear weapons, cars, the written word, eating beef, washing machines, accounting, internal combustion engines, astronomy, and the internet.

Once an innovation gets out there, it can be slowed, or even paused, but it *always* gets adopted if people really want it. It's human nature.

From: Ben   10:08 PM
To: Paul
Subject: We R all SMAC junkies

That doesn't really answer the question: Why don't we just turn it off?

It's because we're addicted, right? And because we like it too much? It's too useful. We literally couldn't live or work without it right now. It's too ingrained in our lives. Too profitable! The economic — and political — power of the FAANGM vendors is at an all-time high.

Because it's impossible to live today and not have a phone and not be in touch and not be on the grid and not scratch the itch of wanting to know what's going on in the big world and/or your small world.

Because it's impossible to sit in a café or on a bus and just stare out the window and not use your thumb. Or to sit in a meeting and concentrate.

It's impossible to just look up and not down.

It's impossible to think something or do something and not want to let everyone know what you thought. Or what you did. It's impossible to just be, rather than be online.

It's impossible to behave like people did for millennia, up until 2004. Right?

From: Paul   10:22 PM
To: Ben
Subject: Burnin' down the house

I think that's exactly right. It's "digital fentanyl." This "thing" — all of technology, social media, endless connectivity, the whole Monster — is a tech innovation that has also become a mass addiction.

None of those things can ever be simply switched off because, honestly, we don't want to!

Ask someone on opioids. Ask someone who smokes cigarettes. Tech gives us the dopamine hits that make it probably even harder to quit.

Ask someone whose stock portfolio is up if they want this off. Ask your kids if they want to stop texting and posting on TikTok.

Ask every aunt, uncle, grandparent in the world if they want to know less about their family members. They don't want Facebook turned off.

Ask an LGBTQ kid in a country where it's illegal if they want their only connection to others to be switched off. No way.

The pandemic proved that we will NEVER turn everything off. We all *needed* tech to stay connected and informed and — for the lucky ones — employed.

Saying, "the hell with it," and turning everything off is like saying we shouldn't use fire to warm ourselves or cook food because fire can burn the house down.

You don't stop using fire; you learn how to use it better. You invent the fireplace. You invent the fire extinguisher. You invent the fire department. You invent the electric generator. You learn to tame fire.

From: Ben   10:39 PM
To: Paul
Subject: Last year's model

What I suspect will happen though is this: Social media *won't be turned off* — not by politicians, not by us; people will *just drift off*. The fashion cycle will turn, and social media will become last year's model.

The next "thing" will come along, and social media will go the way of all things — cave painting, real tennis, opera, music hall, radio, movies, newspapers, books, TV — in becoming less cool and less interesting and a small flatlining niche that's far removed from the mainstream of where the action is.

People will drift off and find other things to do. The uptick in the use of Yondr bags at concerts may be a lead indicator of a changing attitude to listening to, rather than documenting, a musical experience. And then, 20 years after the fact, politicians will act — as they did recently with the "cookie," regulating its use decades after its introduction.

From: Paul   10:45 PM
To: Ben
Subject: "Just Say No" also won't work on digital fentanyl

Maybe. But we don't drift away from loads of things we know are bad for us. Look at smoking. It's been on the decline for decades, but 40 million Americans still smoke cigarettes.

That decline, which is great, didn't happen on its own. It was driven in large part by regulation around access, advertising, tax rates, and public education.

All that, plus, well, the vast majority of smokers die early (and it's not pretty).

Same with seat belts in cars. We had car regulations, education campaigns, etc., but usage really changed when the enforcement laws changed.

If tech use is more like a *real addiction* — not just a fad or passing interest — this isn't going to just stop on its own. The pandemic gave us all a sense of permission to "use" even more.

Some may kick, turn it off, go dark, but most of us won't. Humans need help — through regulations as well as changing social conventions.

From: Ben
To: Paul
Subject: We're doomed

10:57 PM

Will regulation "save" us? It could do, but I don't find the "break 'em up" argument terribly convincing.

If Facebook is broken up, why will that stop Instagram or Snap or TikTok or A.N. Other platform from monetizing our Code Halos?

Well-intentioned as they are, I don't think arguments like those from Larry Sanger, who's trying to get agreement for a Declaration of Digital Independence, will have much chance of changing the weather.

Maybe I'm wrong.

Maybe things like Solid, where Tim Berners-Lee is trying to re-reinvent the World Wide Web, will give us the control that allows us to keep on sharing the cat videos but not be the modern equivalent of the dispossessed indigenous people of the Andes.

Maybe . . .

But I doubt it.

From: Paul   11:35 PM
To: Ben
Subject: The battle isn't over!

For once, I'm a bit more optimistic than you are. Technology *is* hurting us, which is why we're writing this book called *Monster*, but it's also helping, which is why *Monster* is really a tough-love letter. We want to make tech healthy!

I come back to: In the arc of history, these things have *just* been invented, released into the world, and now we are starting to try to figure out what it means, what do we like, what do we not like.

Regulation in all forms — access, tax, advertising, anti-trust, privacy, labor laws, and so much more — haven't even really been deployed to tame the Monster, but they will.

Also, the answer probably isn't just in regulations from governments, companies, and regulatory agencies.

There is a big role for self-regulation via social norms and conventions that we have stepped away from over the last 50 years.

From: Ben   11:38 PM
To: Paul
Subject: Make it stop

I don't know. I really don't.

Why don't we just turn it off?

Why **can't** we just turn it off?

From: Paul   11:45 PM
To: Ben
Subject: Make it stop, or slow it down?

"Can't we just turn it off?" may sound audacious, but it's no more or less crazy to think we can return to some of the practices and conventions we *know* pull us together — spirituality, mindfulness, social interdependence, even compassion.

Those are in us as much as the need to keep swiping and posting, I think. And more likely than "off."

From: Ben   11:46 PM
To: Paul
Subject: Off . . .

Turn it off. . . .

From: Paul   11:48 PM
To: Ben
Subject: Aristotle was right

Who can turn it off???

There is no Big Monster Off-Switch.

Aristotle even noted that the gods can't unbreak an egg. We can't un-invent something.

From: Ben   11:53 PM
To: Paul
Subject: Zuboff is right

Off. Off. *(Zub)off*.

From: Paul   11:59 PM
To: Ben
Subject: Hell is other people on social media

I wish.

But it ain't happening, so we have to get smarter, and kinder. Or it *will* go off . . . for all of us, all at once.

8 POSTFACE

A nd lastly, a word on the title of this book.

As all budding playwrights, screenwriters, and authors are taught in writing school, there are said to be only seven story archetypes, around which every work of fiction — ancient or modern — revolves. When we re-stumbled across this fact, we were naturally intrigued to check the list out. In reverse order, these stories are:

- Rebirth

- Tragedy

- Comedy

- Voyage and Return

- The Quest

- Rags to Riches

And . . .

- **Overcoming the Monster**

At that point, you could have knocked us down with a feather. It seemed like kismet. That what we had been agonizing over for months and months — the theme and title of our manuscript — that felt to us like a completely contemporary, even futuristic, concept was, in fact, the oldest story known to mankind. *Frankenstein, Perseus, Theseus, Beowulf, Dracula, The War of the Worlds, Nicholas Nickleby, The Guns of Navarone, The Magnificent Seven, Star Wars* — all of these famous stories (and many more) rotate around this simple but elemental idea.

When we regained our composure (yes, some drink was taken), we couldn't help but feel that in this light, our latest battle to overcome the Monster, or at least tame it, was justified and right. Even *important.*

Perhaps, we reflected, mankind is doomed to forever face new monsters, the monsters being in the end simply expressions of our monstrous natures. Perhaps, even though modern technology seems different and unprecedented, it isn't. It's just the same challenges we've always faced — to make a living, to be kind, to love, to laugh, to survive — wrapped up in different packaging.

Perhaps, after all, this monster — the tech monster — is no different from any other monster that mankind has faced down through the eons. In time, we suppose, we will see.

We hope, ultimately though, that the thoughts we have shared in this book play some small role in taming this latest version of a monster that walks amidst us. A monster that will shape our jobs, lives, and very future.

Notes

1. Have we created a monster?

1. Gartner. (2019). Gartner says global IT spending to reach $3.8 trillion in 2019 (January 28). https://www.gartner.com/en/newsroom/press-releases/2019-01-28-gartner-says-global-it-spending-to-reach--3-8-trillio.

2. The average non-fiction book runs around 65,000 or 70,000 words. We kept this to about half that by spending days, nights, and weekends over the past several years grinding the ideas down as far as we could, inspired by the old adage, "If I'd had more time, I would have written you a shorter letter." You'll no doubt conclude that every chapter in our short book could be its own long book, but we've tried to take out all the extra stuffing and clichés, so you can read it in a couple of hours. (You're welcome!)

2. Machines

1. Our 2017 book *What to Do When Machines Do Everything* covers this ground at much greater length. This is simply an intro for those of you who haven't read the book (shame on you) and a refresher for those of you who have (you're the tops).

2. The British Council. (2014), 80 moments that shaped the world. https://www.britishcouncil.org/sites/default/files/80-moments-report.pdf.

3. YouTube. (2017). The sounds of IBM: IBM Q (December 14). www.youtube.com/watch?v=o-FyH2A7Ed0.

4. D-Wave. Introduction to the D-Wave quantum hardware. https://www.dwavesys.com/tutorials/background-reading-series/introduction-d-wave-quantum-hardware.

5. Arute, F., Arya, K., Babbush, R. et al. (2019). Quantum supremacy using a programmable superconducting processor. *Nature*. https:// www.nature.com/articles/s41586-019-1666-5.

6. Waters, R. (2019). The billion-dollar bet to reach human-level AI. *Financial Times*. https://www.ft.com/content/c96e43be-b4df-11e9-8cb2-799a3a8cf37b.

7. Gunther McGrath, R. (2019). The pace of technology adoption is speeding up. *Harvard Business Review* (September 25). https://hbr.org/2013/11/the-pace-of-technology-adoption-is-speeding-up.

8. Silva, D. (2009). Internet has only just begun, say founders. *PhysOrg*. https://phys.org/news/2009-04-internet-begun-founders.html.

9. Greenberg, A. (2017). Hacking North Korea is easy. Its nukes? Not so much. *Wired*. https://www.wired.com/story/cyberattack-north-korea-nukes.

10. Aaron Levie on Twitter. https://twitter.com/levie/status/547234465198526464?lang=en.

11. Gan, N. (2020). China is installing surveillance cameras outside people's front doors . . . and sometimes inside their homes. CNN. https://www.cnn.com/2020/04/27/asia/cctv-cameras-china-hnk-intl/index.html.

12. Nichols, G. (2020). As lockdowns ease, a new surveillance reality awaits. ZDNet. https://www.zdnet.com/article/as-lockdowns-ease-a-new-surveillance-reality-awaits.

13. Chin, M. (2020). Exam anxiety: How remote test-proctoring is creeping students out. *The Verge*. https://www.theverge.com/2020/4/29/21232777/examity-remote-test-proctoring-online-class-education.

14. Waddell, K. (2018). The U.S-China race for quantum dominance. Axios. https://www.axios.com/us-chinese-race-for-quantum-dominance-63f55bc2-45b5-438f-ae5c-a29b32a42642.html.

15. Alan Boyle. (2018). Trump signs legislation to boost quantum computing research with $1.2 billion." *GeekWire* (December 21). www.geekwire.com/2018/trump-signs-legislation-back-quantum-computing-research-1-2-billion.

16. deSolla Price, L. (2019). Scientists thought the first atom bomb tested might destroy the world. FactMyth. http://factmyth.com/factoids/scientists-thought-the-first-atom-bomb-tested-might-destroy-the-world.

3. Capital

1. Eavis, P. and Lohr S. (2020). Big Tech's domination of business reaches new heights. *New York Times*. https://www.nytimes.com/2020/08/19/technology/big-tech-business-domination.html.

2. Piketty, T. (2014). *Capital in the Twenty-First Century*. Cambridge, MA: The Belknap Press of Harvard University Press.

3. Unfortunate historical language alert: "Men" as defined by Thomas Jefferson meant white men from Northern Europe and not "mankind."

4. Picchi, A. (2018). Inequality is worsening and could hit U.S. credit rating: Moody's. *CBS News* (October 8).

5. Santnes, S. "Does Guaranteeing a Basic Income Reduce Income Inequality?" www.scottsantens.com/does-basic-income-reduce-income-inequality-gini. World Inequality Lab. (2018). World Inequality Report 2018. http://wir2018.wid.world.

6. Picchi, A. (2018). Inequality is worsening and could hit U.S. credit rating: Moody's. *CBS News* (October 8). Obviously, a country can have a low Gini score and not be economically healthy; e.g., poorer countries like Belarus have much lower scores.

7. Gunther McGrath, R. (2019). The pace of technology adoption is speeding up. *Harvard Business Review*. https://hbr.org/2013/11/the-pace-of-technology-adoption-is-speeding-up.

8. Florida, R. (2015). Income inequality leads to less happy people. Bloomberg CityLab. https://www.citylab.com/equity/2015/12/income-inequality-makes-people-unhappy/416268.

9. Zuboff, S. (2019). *The Age of Surveillance Capitalism: The Fight for a Human Future at the New Frontier of Power*. New York: PublicAffairs.

10. Stop-Covid.tech. 13 things tech companies can do to fight coronavirus. https://stop-covid.tech.

11. Stack Exchange. Why is the UK called Airstrip One? https://literature.stackexchange.com/questions/71/why-is-the-uk-called-airstrip-one. Chu, C.M., Andrews, C.H., and Gledhill, A.W.(1950). Influenza in 1948–1949. *Bulletin of the World Health Organization* 3 (May): 187–214. https://www.ncbi.nlm.nih.gov/pmc/articles/PMC2553936/pdf/bullwho00643-0003.pdf.

12. Kerry, C.F. (2018). Why protecting privacy is a losing game today — and how to change the game. Brookings. https://www.brookings.edu/research/why-protecting-privacy-is-a-losing-game-today-and-how-to-change-the-game.

13. Parrish, S. Gates' Law: How progress compounds and why it matters. FS. https://fs.blog/2019/05/gates-law.

14. Thompson, D. (2012). The economic history of the last 2,000 years in 1 little graph. *The Atlantic*. https://www.theatlantic.com/business/archive/2012/06/the-economic-history-of-the-last-2-000-years-in-1-little-graph/258676. Infogram. Share of world GDP throughout history. https://infogram.com/share-of-world-gdp-throughout-history-1gjk92e6yjwqm16.

15. Parks, B., Bluhm, R., Dreher, A., et al. (2018). Belt and road projects direct Chinese investment to all corners of the globe. What are the local impacts? *Washington Post* (September 11). *China Daily*. (2019). Belt, road markets drive strong growth in exports, imports (May 10). https://www.washingtonpost.com/news/monkey-cage/wp/2018/09/11/belt-and-road-projects-direct-chinese-investment-to-all-corners-of-the-globe-what-are-the-local-impacts/.

16. CNBC. (2018). Where is the funding for a $26 trillion initiative coming from? (March 6). https://www.cnbc.com/advertorial/2018/03/06/where-is-the-funding-for-a-26-trillion-initiative-coming-from.html.

17. Hillman, J. (2019). Five myths about China's Belt and Road Initiative. *Washington Post*. https://www.washingtonpost.com/outlook/five-myths/five-myths-about-chinas-belt-and-road-initiative/2019/05/30/d6870958-8223-11e9-bce7-40b4105f7ca0_story.html.

18. Belt and Road News. (2019). Cold War for artificial intelligence supremacy looming (January 11). https://www.beltandroad.news/2019/01/11/cold-war-for-artificial-intelligence-supremacy-looming.

19. Frank, M., Roehrig, P., and Pring, B. (2017). *What to Do When Machines Do Everything: How to Get Ahead in a World of AI, Algorithms, Bots, and Big Data*, Hoboken, NJ: John Wiley & Sons.

20. Wang, S. and Lahiri, T. (2019). A future AI park in Malaysia shows how criticism is changing China's foreign investment. *Quartz*. https://qz.com/1602194/an-ai-park-in-malaysia-shows-chinas-belt-and-road-is-evolving.

21. Ibid.

22. Wijeratne, D., Rathbone, M., and Wong, G. (2019). A strategist's guide to China's Belt and Road Initiative. *strategy+business*. https://www.strategy-business.com/feature/A-Strategists-Guide-to-Chinas-Belt-and-Road-Initiative.

23. Araya, D. (2019). China's grand strategy. *Forbes*. https://www.forbes.com/sites/danielaraya/2019/01/14/chinas-grand-strategy/#61b7c6c01f18.

24. Tracy, B. (2018). China assigns every citizen a "Social Credit Score" to identify who is and isn't trustworthy. CBS. https://newyork.cbslocal.com/2018/04/24/china-assigns-every-citizen-a-social-credit-score-to-identify-who-is-and-isnt-trustworthy.

25. Doffman, Z. (2020). Black Lives Matter: U.S. protesters tracked by secretive phone location technology. *Forbes* (June 26).

26. Leverett F.L. and Sprinkle R. (2017). China steps up as U.S. steps back from global leadership. *The Conversation*. http://theconversation.com/china-steps-up-as-us-steps-back-from-global-leadership-70962.

27. Frankie Goes to Hollywood. (1984). "Two tribes." ZZT Records.

28. Maçães, B. (2018). What you need to know to understand the Belt and Road. World Economic Forum. https://www.weforum.org/agenda/2018/12/what-you-need-to-know-one-belt-one-road.

29. Roehrig, P. (2019). The internet is broken. But we can't just repair it — we need to rebuild it. *Quartz*. https://qz.com/1631441/the-splinternet-exists-and-we-need-to-fix-it.

4. Psychology

1. Hasel, T. (2020). Zoom falls 11% after CEO apologizes for security lapses, says daily users spiked to 200 million in March. CNBC. https://www.cnbc.com/2020/04/02/zoom-ceo-apologizes-for-security-issues-users-spike-to-200-million.html.

2. The total number of people living with depression — now more than 322 million across the globe — continues to grow. See: World Health Organization. (2017). Depression and other common mental disorders. https://www.who.int/publications-detail/depression-global-health-estimates. The suicide rate among the young in the US was relatively stable for years until the late 2000s — roughly when smartphones became nearly ubiquitous — and this has increased horribly since then. Teens who spend hours on their smartphones every day — seemingly most of them — are a terrifying 71% more likely to have higher risk factors for suicide. See: Lulu Garcia-Navarro. (2017). The risk of teen depression and suicide is linked to smartphone use, study says. NPR (December 17). https://www.npr.org/2017/12/17/571443683/the-call-in-teens-and-depression. A growing body of evidence links time on social media preceding unhappiness. See: Jean Twenge. (2018). Are smartphones causing more teen suicides? *The Guardian* (May 24). https://www.theguardian.com/society/2018/may/24/smartphone-teen-suicide-mental-health-depression.

3. The University of Alabama. The 'More Doctors Smoke Camels' campaign. https://csts.ua.edu/ama/more-doctors-smoke-camels.

4. Jack Shephard. (2019). "Keith Richards says heroin is easier to kick than cigarettes. *The Guardian* (February 14). https://www.independent.co.uk/arts-entertainment/music/news/keith-richards-heroin-cigarettes-alcohol-drugs-addiction-rolling-stones-a8778861.html.

5. One great summary of the cognitive revolution is by Yuval Noah Harari in his book *Sapiens: A Brief History of Humankind*, Harper, 2015.

6. These examples and many more, along with the data and psychological reasoning for these biases, can be found in Daniel Kahneman's *Thinking, Fast and Slow*, Farrar, Straus and Giroux, 2015.

7. Kanter, J. (2019). Facebook is to democracy what smoking is to your health, say technical experts. *Business Insider*. https://Jan. 23, 2019, www.businessinsider.com/facebook-is-a-damaging-to-your-health-as-cigarettes-say-tech-experts-2019-1.

8. Dodds, L. (2019). "We lost control of our creations": The Silicon Vally heretic on a mission to make Big Tech repent. *The Telegraph*. https://www.telegraph.co.uk/technology/2019/05/10/lost-control-creations-silicon-valley-heretic-mission-make-big.

9. Amer, K. and Noujaim, J., dirs. (2019). *The Great Hack*. The Othrs. https://www.netflix.com/title/80117542.

10. Scott, M. (2018). Cambridge Analytica helped "cheat" Brexit vote and US election, claims whistleblower. *Politico*. https://www.politico.eu/article/cambridge-analytica-chris-wylie-brexit-trump-britain-data-protection-privacy-facebook.

11. Hayes, S.C. (2014). The unexpected way that new technology makes us unhappy. *Psychology Today*. https://www.psychologytoday.com/us/blog/get-out-your-mind/201409/the-unexpected-way-new-technology-makes-us-unhappy.

12. Annunziata, M. (2019). The great cognitive depression. *Forbes*. https://www.forbes.com/sites/marcoannunziata/2019/01/11/the-great-cognitive-depression/#7036bb2774c1.

13. List of best-selling mobile phones. Wikipedia. https://en.wikipedia.org/wiki/List_of_best-selling_mobile_phones.

14. Eadicicco, L., Peckham, M., Fitzpatrick, A. et al. (2016). The 50 most influential gadgets of all time. *Time*. https://time.com/4309573/most-influential-gadgets/.

15. Taibbi, M. (2010). The great American bubble machine. *Rolling Stone*. https://www.rollingstone.com/politics/politics-news/the-great-american-bubble-machine-195229.

16. Ritchie, H. and Roser, M. (2018). Mental health. Our World in Data. https://ourworldindata.org/mental-health.

17. Brown, M. (2019). Drug abuse is a main driver behind America's rising rate of "deaths of despair." *Deseret News*. https://www.deseret.com/indepth/2019/9/5/20851638/drug-abuse-is-a-main-driver-behind-americas-rising-rate-of-deaths-of-despair.

18. The Rolling Stones. (1969. "Gimme Shelter" lyrics, *Let It Bleed*. Decca Records.

5. Society

1. Daugherty, O. (2018). People are slashing tires on self-driving vehicles in Arizona. *The Hill*. https://thehill.com/policy/technology/423374-people-are-slashing-tires-on-self-driving-vehicles-in-arizona. Durden, T. (2019). Tires slashed, guns pulled on self-driving cars as Arizona residents revolt. ZeroHedge (January 1). https://www.npr.org/2019/01/02/681752256/why-phoenix-area-residents-are-attacking-waymos-self-driving-fleet.

2. Rushkoff, D. (2016). *Throwing Rocks at the Google Bus: How Growth Became the Enemy of Prosperity*. Portfolio/Penguin.

3. Hapsoro, M. (2019). South Korea's taxi drivers keep setting themselves on fire to protest a new ridehailing service. *Vice*. https://www.vice.com/en_asia/article/qvyzn5/south-korea-uber-kakao-taxi-protests-self-immolate.

4. Bensaâdoune, N. (2018). If you want to understand the gilets jaunes, get out of Paris. *The Guardian*. https://www.theguardian.com/commentisfree/2018/dec/12/gilets-jaunes-paris-protesters-france.

5. ENIAC entry. Wikipedia. https://en.wikipedia.org/wiki/ENIAC.

6. Gerstell, G.S. (2019). I work for N.S.A. We cannot afford to lose the Digital Revolution, *New York Times*. https://www.nytimes.com/2019/09/10/opinion/nsa-privacy.html.

7. Shrinkman, P.D. (2013). Former CIA Director: Cyber attack game changers comparable to Hiroshima. *U.S. News*. https://www.usnews.com/news/articles/2013/02/20/former-cia-director-cyber-attack-game-changers-comparable-to-hiroshima. Greenberg, A. (2019). The Wired guide to cyberwar. *Wired*. https://www.wired.com/story/cyberwar-guide.

8. Giles, M. (2019). Triton is the world's most murderous malware, and it's spreading. *MIT Technology Review*.

9. Greenberg, A. (2017). How an entire nation became Russia's test lab for cyberwar. *Wired*. https://www.wired.com/story/russian-hackers-attack-ukraine.

10. Gerald R. Ford aircraft carrier entry on Wikipedia. https://en.wikipedia.org/wiki/Gerald_R._Ford-class_aircraft_carrier. Boyd, A. (2019). Trump's 2020 budget requests about $11 billion for cyber defense and operations. Nextgov. https://www.nextgov.com/cybersecurity/2019/03/trumps-2020-budget-requests-about-11-billion-cyber-defense-and-operations/155445.

11. Manson, K. (2018). Robot soldiers, stealth jets and drone armies: The future of war. *Financial Times*. https://www.ft.com/content/442de9aa-e7a0-11e8-8a85-04b8afea6ea3.

6. A manifesto for taming the Monster

1. A tip o' the hat to Douglas Adams for just one of the many memorable lines of *The Hitchhiker's Guide to the Galaxy*, Pan Books, 1979.

2. Roberts, J.J. (2016). Tech industry wins big in Supreme Court patent ruling. (June 20). *Alice Corporation v. CLS Bank International*. 134 S. Ct. 2347 (2014). http://fortune.com/2016/06/20/tech-industry-wins-big-in-supreme-court-patent-ruling.

3. The Alice decision and its fallout in the U.S. (2018). *Morningside Translations* (September 25). https://www.morningtrans.com/the-alice-decision-and-its-fallout-in-the-u-s/.

4. Rappeport, A., Schreuer, M., Tankersley, J., and Singer. N. (2018). Europe's planned digital tax heightens tensions with U.S. *New York Times*. https://www.nytimes.com/2018/03/19/us/politics/europe-digital-tax-trade.html.

5. Finance chiefs warn on Big Tech's shift to banking. (2018). *Financial Times* (February 4). https://www.ft.com/content/d9b3d79e-0995-11e8-8eb7-42f857ea9f09.

6. Bach, N. (2018). Amazon is reportedly in talks with JP Morgan to create checking accounts. http://fortune.com/2018/03/05/amazon-bank-jp-morgan-checking-account.

7. Gurman, M. and Surane, J. (2019). Apple debuts titanium credit card with Goldman, Mastercard. *Bloomberg*. https://www.bloomberg.com/news/articles/2019-03-25/apple-announces-apple-credit-card-with-goldman-sachs.

8. Henley, J. (2017). Uber to shut down Denmark operation of new taxi laws. *The Guardian*. https://www.theguardian.com/technology/2017/mar/28/uber-to-shut-down-denmark-operation-over-new-taxi-laws.

9. Eavis, P. and Lohr, S. (2020). Big Tech's domination of business reaches new heights. *New York Times*. https://www.nytimes.com/2020/08/19/technology/big-tech-business-domination.html.

10. Speaking of fines, Facebook was potentially liable for a $40,000 fine for each of the 50 million users who had their data (allegedly) snorkeled away into Cambridge Analytica's (alleged, potentially) election-shaping AI engines. That could have been a $2 trillion potential exposure. Powell, K. (2018). Facebook could face $2 trillion in fines if found in violation of data scandal. YourEDM. https://www.youredm.com/2018/03/20/facebook-could-face-2-trillion-in-fines-if-found-in-violation-of-data-scandal.

11. What is GDPR, the EU's new data protection law. https://gdpr.eu/what-is-gdpr.

12. GDPR fines and penalties. www.gdpr.eu/compliance/fines-and-penalties.

13. Clerkin, B. (2017). Who's the boss? Congress eyes autonomous vehicle rules that would wipe out states' oversight of technology. DMV News.

14. Bellon, T. (2017). Congress is trying to quickly pass a self-driving car law — and states are freaking out. *Business Insider*. https://www.businessinsider.com/r-us-push-for-self-driving-law-exposes-regulatory-divide-2017-9.

15. Meyer, D. (2018). Europe just dropped a major tax bombshell to get U.S. tech giants to pay up. *Fortune*. http://fortune.com/2018/03/21/europe-tech-firms-tax-bombshell.

16. Ronald Reagan Presidential Foundation & Institute. https://www.reaganfoundation.org/ronald-reagan/reagan-quotes-speeches/inaugural-address-2. Stone, Oliver. (1987). *Wall Street*. Twentieth Century Fox.

17. Economist Joseph Schumpeter posited that major economic changes require that the existing structure be destroyed from within as the new system is created. See the "creative destruction" entry on Wikipedia. https://en.wikipedia.org/wiki/Creative_destruction.

18. The Homebrew Computer Club was a hobbyist group that helped spark the PC revolution. Members included Steve Jobs, Steve Wozniak. and others who ultimately founded several microcomputing companies. See the Homebrew Computer Club entry in Wikipedia. https://en.wikipedia.org/wiki/Homebrew_Computer_Club.

19. O'Shea, L. (2018). Time to cut ties with the digital oligarchs and rewire the web. *The Guardian*. https://www.theguardian.com/commentisfree/2018/mar/20/digital-oligarchs-rewire-web-facebook-scandal.

20. SpyCloud (2020). Annual Credential Exposure Report 2020. https://spycloud.com/2020-annual-credential-exposure-report.

21. Shape Security. 2018 Credential Spill Report. http://info.shapesecurity.com/rs/935-ZAM-778/images/Shape_Credential_Spill_Report_2018.pdf.

22. Morgan, S. (2018). Global cybercrime damages predicted to reach $6 trillion annually by 2021. *Cybercrime Magazine*. https://cybersecurityventures.com/cybercrime-damages-6-trillion-by-2021.

23. Kafka, P. (2020). Facebook is like sugar — too much is bad for you, says a top Facebook exec. *Recode*. https://www.vox.com/recode/2020/1/7/21056094/facebook-sugar-regulation-memo-trump-2016-bosworth.

24. DiNicolantonio, J.J., O'Keefe, J.H., and WIlson, W.L. (2018). Sugar addiction: Is it real? A narrative review. *British Journal of Sports Medicine* 52 (14). https://bjsm.bmj.com/content/52/14/910.

25. Fox, M. (2013). Heavy burden: Obesity may be even deadlier than thought. *NBC News*. https://www.nbcnews.com/healthmain/heavy-burden-obesity-may-be-even-deadlier-thought-6C10930019.

26. If establishing healthy limits is just too difficult, then it's like any other addiction (and there can be true digital addictions). There is help available, and we both hope you seek it out and do the hard work to get clean. Trying to get out from under it on your own is like trying to perform surgery on yourself: theoretically possible, but unlikely to work out well.

27. "History of Accounting." Wikipedia entry. https://en.wikipedia.org/wiki/History_of_accounting; Stock Market entry in Wikipedia, https://en.wikipedia.org/wiki/Stock_market.

28. "Failures to conceive of zero-price markets as antitrust 'markets' indicate how fundamentally zero prices challenge traditional theories and analytical frameworks." Newman, J.M. (2015). Antitrust in zero price markets: Foundations. *University of Pennsylvania Law Review* 164. https://scholarship.law.upenn.edu/cgi/viewcontent.cgi?article=9504&context=penn_law_review.

Acknowledgments

First and foremost, we'd like to acknowledge the support and contributions of Malcolm Frank, co-author for our two preceding books. Even though he did not put pen directly to paper for this one, all of these ideas have been significantly informed and shaped by many (many) workshop sessions with him. His voice and counsel remained at the top of our minds as we wrestled these ideas into words.

We owe an incalculable debt to countless clients, company leaders, thought leaders, journalists, and academics who have helped influence our ideas and view of the world. The leadership teams of Cognizant — particularly CEO Brian Humphries, CMO Gaurav Chand, and former CEO Francisco D'Souza — have continued to provide indefatigable support as we explore bold ideas to help our clients and fellow associates. The Cognizant Center for the Future of Work team members are trusted colleagues and friends to both of us, and their input, trust, and confidence continue to help push us forward. We'd particularly like to thank Robert H. Brown for his help stress-testing the concepts of the overall piece. Alan Alper has been with us for the entire journey, providing world-class editing, thoughtful content guidance, friendship, an impressive knowledge of 1980s music, and endless patience.

Finally, and most importantly, none of this — literally none of this — would be possible or fun without the insight, ideas, trust, and love of our families. Thank you for being so understanding and encouraging of us through the innumerable days and nights.

About the Authors

Paul Roehrig

Paul Roehrig is the Global Head of Strategy and Growth for Cognizant Digital Business. He is the co-founder and former Global Managing Director of the Center for the Future of Work at Cognizant and is a member of Cognizant's Executive Leadership Team. He is also — along with Malcolm Frank and Ben Pring — a co-author of the best-selling and award-winning books *What to Do When Machines Do Everything: How to Get Ahead in a World of AI, Algorithms, Bots, and Big Data* and *Code Halos: How the Digital Lives of People, Things, and Organizations Are Changing the Rules of Business.*

Prior to joining Cognizant, Paul was a Principal Analyst at Forrester Research, where he researched, wrote, and consulted extensively on challenges and opportunities related to business and technology services. In 2008, Paul was named one of the top services analysts globally by the Institute of Industry Analyst Relations. He also held operational positions in designing and implementing global technology services programs for customers from a variety of industries for Hewlett-Packard.

Paul holds a PhD from Syracuse University and a journalism degree from the University of Florida. He is based in Washington, DC.

Ben Pring

Ben Pring is the head of thought leadership at
Cognizant, co-founded and leads Cognizant's
Center for the Future of Work, and is a
member of Cognizant's Executive Leadership
Team. Ben is a co-author of the best-selling
and award-winning books, *What to Do When
Machines Do Everything* (2017) and *Code
Halos: How the Digital Lives of People, Things,
and Organizations Are Changing the Rules of
Business* (2014).

Ben sits on the advisory board of the
Labor and Work Life program at Harvard Law School. In 2018
Ben was a Bilderberg Meeting participant. Ben was named as one
of 30 management thinkers to watch in 2020 by Thinkers 50. He
was recently named a leading influencer on the future of work by
Onalytica.

Ben joined Cognizant in 2011, from Gartner, where he spent 15 years
researching and advising on areas such as Cloud Computing and
Global Sourcing. In 2007, Ben won Gartner's prestigious Thought
Leader Award. Prior to Gartner, Ben worked for a number
of consulting companies including Coopers and Lybrand.

Based near Boston since 2000, Ben graduated with a degree in
Philosophy from Manchester University in the UK.

Index